COLD EARTH

COLD EARTH

SARAH MOSS

COUNTERPOINT
BERKELEY

Originally published in English by Granta Publications under the title Cold Earth copyright © Sara Moss, 2009.

First published in Great Britain by Granta Books 2009.

Library of Congress Cataloging-in-Publication Data
Moss, Sarah.
Cold earth : a novel / by Sarah Moss.
p. cm.
Summary: A team of six archaeologists from the United States, England, and Scotland assembles at the beginning of the Arctic summer to unearth traces of the lost Viking settlements in Greenland. But as they sink into uneasy domesticity, there is news of an epidemic back home, and their communications with the outside world fall away. Facing a Greenland winter for which they are hopelessly ill-equipped, Nina, Ruth, Catriona, Jim, Ben, and Yianni, knowing that their missives may never reach their loved ones, write final letters home. These letters make up the narrative of Cold Earth, with each section of the book comprised of one character's first-person perspective in letter form. In this exceptional and haunting debut novel, Moss weaves a rich tapestry of personal narrative, history, love, grief, and naked survival. Cold Earth is both a heart-pounding thriller and a highly sophisticated novel of ideas.

ISBN 978-1-58243-579-4
1. Archaeologists—Greenland—Fiction. 2. Survival after airplane accidents, shipwrecks, etc.—Greenland—Fiction. 3. Greenland—Fiction. I. Title.

PR6113.O87C65 2010
823'.92—dc22
2009052549

Cover design by Mark Swan and Domini Dragoone
Interior design by M Rules
Printed in the United States of America

COUNTERPOINT
2117 Fourth Street
Suite D
Berkeley, CA 94710

www.counterpointpress.com

Distributed by Publishers Group West

10 9 8 7 6 5 4 3 2 1

For Kathy, with thanks.

NINA

I couldn't sleep, the first night here. It was partly excitement, the relief of finally being in Greenland, and partly the light. I think I'd expected midnight sun to be obviously exotic, but it's only slightly different, as if the sun has looked away, and you know what I'm like about sleeping in the day. It felt as if I ought to get up and do some work. I lay there, hot in my sleeping bag, eyes prickling as if I'd got sand in them, and stared at the tent until it seemed to stare back. You were right, I should have got round to a trial run in the park. It's not as easy as it ought to be, putting up tents. The sides looked increasingly lop-sided as the night wore on. I did check for stones before I spread the groundsheet, but the longer I lay there the more stones I seemed to be lying on, and after a while it was clear to me that one of them had a long straight edge and was therefore man-made and in all probability a gravestone. I tried not to think about it. I closed my eyes and counted my breathing, remembering the voice of that Scottish woman on the relaxation tape. Let your fingers soften. Feel your wrists loosen. Shame she sounds like a headmistress moonlighting as a phone-sex worker. Let the tension flow out of your shoulders.

It takes a long time to set fire to a church. Their voices come up the valley in the dark, sniggering and swearing like schoolboys. The stones will never burn, and the turf roof takes a while to dry out, but at last something begins to crackle and pale smoke rises in the dark sky. The men holler and leap as the windows glow orange and the dancing light shows twisted shapes on the ground between the church and the river. A cry echoes over the water from the waiting ship. The last few deaths were not quick or easy. I shiver, huddled on damp turf behind a rock, and try not to think about what will happen to the women on board when the sailors return to the boat. Screams carry a long way across the quiet sea.

I sat up, the tension flowing back. Sometimes it's better to be stressed and awake. Reality is bad enough without having to bear unearthly presences as well, though the headlines I saw at Heathrow open the contested border between one's worst imaginings and *les actualités*. It's usually a mistake to think about the news, I know, but worse when travelling, and a particularly bad idea to think about people you love and the news at the same time when you're nowhere near either of them. There's something about dislocation that makes the news seem horribly probable in a way that it doesn't at home. I want to admit now that I don't really like travelling. I never have. I've been pretending to be brave and sophisticated these last four years, but honestly a cottage in Cornwall, by train to avoid possible pile-ups on the M4, is probably about as far as I'd go if left to myself. Which I do not wish to be.

I do like being a well-travelled person. It sort of makes all those trips worthwhile, the status, and I know I'm good at planning, but in fact I've always wished that meticulous organisation would displace the obligation actually to go. Maybe I should work for one of those bespoke travel agencies with the art nouveau fonts, putting together exquisite tours for people untrammelled by cost. I always worry most the night before going to America. I like the idea of America, of people who are able to entertain the possibility that strangers might be worth talking to and that there are circumstances in which one might reasonably wish to order lunch in a restaurant after 1.30 pm, but with hours to go before take-off the dark side of freedom becomes apparent. I've seriously thought about hijacking planes after changing my mind over the Atlantic. Officer, there's been a terrible misunderstanding. Take me home. Do you know how many serial killers they have roaming the US at any given moment, looking for random strangers such as lost Englishwomen whose partners are busy in some gallery on whom to perpetrate acts of arbitrary violence? Do you know how many European tourists have been shot by American householders for coming to the door to ask directions? Not to mention drive-by shootings. It's insane to risk a perfectly well-ordered life for some whim like going to America. I always plan to like the hotels, bathrooms sanitised for my protection and lots of clean sheets, but a paper banner saying something is 'sanitised' doesn't mean it's clean. How much effort would you put into a bathroom that's going to be used by people you never see? And in the South it's hopeless, I simply couldn't make myself understood. Not even in French. I was too ashamed to tell you that one day I got faint with hunger in that town outside Atlanta because I couldn't identify a shop selling food and all attempts to communicate with waiters ended in mutual incomprehension and painful embarrassment. I gave up and spent the day lying on the bed rationing the last of the jelly beans and reading *Daniel Deronda*. I did better in Greece with no Greek.

I've been happier in Europe but not much. I liked the food, but they kill people on purpose quite often in Corsica, and I worried about bombs all the time in restaurants. Italian roads are dangerous, the Swiss have almost as many guns as the Georgians and even the Danes are given to drink-driving. The real reason I wouldn't go to Russia was the horror of making a fool of myself in a place where I can't even read the alphabet, and I applied for that conference in Rome after you planned the trip to Japan. I am sorry. The only place I really enjoyed was Iceland, so cool and beautiful and safe. Do you remember sitting on that hillside after picking all the blueberries? We could hear only birds and wind, and chocolate didn't melt even though it was August, and then later those German backpackers told us it was thirty-five degrees in London. When I rule the world I'm going to set a maximum midday temperature of the point at which good chocolate makes a noise when you break it.

I thought Greenland would be like Iceland, but I wasn't expecting this to be easy. Honestly. It's not one of those times I get over-excited and then fail to cope with the real thing. The headlines made it worse, and remember it was you who insisted I come when my nerve failed at Heathrow. (I keep wondering how many nerves fail at Heathrow, but it's like a wedding, isn't it, once a person has set off down the aisle or the check-in queue the current of ritual is overpowering.) I can still hear the beating of your heart against my ear and feel the roughness of your cheek against my forehead but I can remember only the words I use for your smell. You stroked my hair and told me I'd like it when I got here, like Australia, but the thing is I didn't like Australia. It was too hot and all the women in Sydney were too elegant. I grant you that neither heat nor elegance is a difficulty in Greenland. I didn't mention the headlines, didn't want you to know that I was thinking about it, but I held you so that now I can still feel your body against mine. Where it belongs. It felt, as it always does feel, an act of violence to walk away from you, there in the departures lounge. I hurt myself when I leave you. If I had looked back I would not be here now. Heat rose behind my eyes. I bit my lip and checked my watch. Hours to go. Claire says trying to keep your eyes open is a surprisingly effective way of getting to sleep. I stared at the pink canvas and waited.

As the flames crackle and the turf roof dries and begins to singe, someone is trying to get out. The sweet smoke of a turf fire rises in the silent valley, and the sheep move nervously towards the river. The priest is trapped in there. I saw him slip in, during the killing, and now I can hear him begging to get out. The glass in the windows – brought from Norway at great expense – has fragmented and fallen onto the soft grass of the churchyard, but I do not think he will get out that way. Flames are reaching up through the window-frames and the timber door is smouldering. He has stopped praying. The noise he is making now rings across the valley and the water in the darkness, and as the flames rise I see a hooded figure in the window, arms stretched out towards the running river and the empty houses. I am sure he sees me, standing in the pasture where I can feel the heat on my face. The smell of burning changes. The stones at the bottom of the bell-tower are blackening and the bell begins to toll softly as the roof falls in and the sky brightens in the east.

No. I opened my eyes again and the pink stared back. Calm thoughts. Sometimes it works to count the good things from the day before and in the day to come, even if the only thing to look forward to is eating. Tomorrow is another day but at least there will be breakfast. I reminded myself that the stones are only a problem because I'm still thin, not yet like my mother, and forgave myself all the things I ate on the way. Crisps at Heathrow, in case it was my last chance until the autumn, foreigners tending to show a strange resistance to the charms of salt and vinegar flavouring. There's a good bakery at Copenhagen airport and I made the most of it, knowing that rural Greenland won't run to bakeries and being fairly sure Yianni wouldn't let me build an oven. I expect you could bury cakes with hot stones on the beach, the way Nic and Mike did with the clams at Rock Point, but of course the egg supply would be difficult. Anyway, I had some nice pastries and a very good chocolate cake. Not too sweet. And then it occurred to me that there wouldn't be ice-cream here either and there was a sort of gelateria, though in the end it looked better than it was. Synthetic flavours in the raspberry, and the chocolate had a lower cocoa content than I do. I had two hours at Nuuk and ate a pizza more out of boredom than anything else, which was all right considering it was in an airport in Greenland. I bet the first tomato landed in Greenland after the Second World War. I'm sure Yianni's better organised than we were but I kept remembering how we ended up living on yoghurt and marzipan in Iceland. Well, there are worse ways to survive.

So I was lying there thinking about food, and wondering what ingredients Yianni had brought and what I could do with them, trying not to think about you or dead people or the news or what time it was and how much sleeping time I'd wasted trying not to think about things. I breathed deeply for a bit and then wondered about getting up and going for a walk, and then something outside the tent made a horrible wrenching, tearing noise right by my ear, as if dead hands were forcing themselves up from the grave. Slowly I turned my head, but whatever it was cast no shadow and I couldn't tell if that was because of the angle of the sun or because it was a supernatural presence. I lay with my joints locked, trying to hear more terrible sounds over the thudding in my ears, and it came again, behind my head. I remembered about the windigo. You were away when I wrote about the windigo, but it was part of that stuff about cannibalism, which has more to do with nineteenth-century travel writing than is quite seemly. The windigo is a monster described to Hudson's Bay Company traders by the local Native Americans. It was once a person who ate human flesh and went mad with the desire for more. You can tell a windigo because it sneaks around camps at night emitting a whistling noise that only the intended victim can hear. I wondered if the people in surrounding tents could hear the flesh being ripped and munched, and if so, whether there was anything I could do to stop it turning to me next. The tent sides quivered in the sun and I held my breath. I wondered how much you would mind if I got eaten by a monster my first night away, and then I thought how much I would mind if you did. Then I started to cry properly and sat up, recklessly alerting all the cannibals and monsters in the valley to my presence. One of them bleated and scampered off, little hooves drumming the turf. An alternative interpretation of available evidence would suggest sheep.

I lay down again and banged my elbow on the probable burial site. The tents are across the river from the church and I'm sure we're not camping on consecrated ground, in which case I decided any burials would be suicides or, more probably in medieval Greenland, secret murder victims. (I know nobody puts memorial stones over secret murder victims – Here Lies the Body, I Dunnit – but that wasn't obvious at two in the morning, OK?) There's a ghost story in one of the sagas William Morris liked about an isolated little bothy at the side of a mountain track, a place for benighted travellers to wait for dawn. Sometimes an angry dead man came out and stabbed everyone while they were asleep. The screams carried down the valley when the wind was in the right direction and the villagers would send for the priest before investigating. I wriggled about, wondering about inscriptions, and started to get cross that Yianni had let me sleep on the grave of a bloodthirsty wraith. I suppose archaeologists have to cope with these things but he knows I don't like them. Being cross is even worse for insomnia than being scared. The sun moved round so hot pink light came straight onto my face and seagulls started shouting at each other. I gave up on sleep, feared that the nearest I'd get to a shower would be a very cold river and wondered if there was anywhere to pee.

After the fire, an old woman comes out from the silent farmstead and walks down the hill to the church. The walls stand, but the roof has gone and smoke is still rising into the pale sky. A wind carries the smoke at a slow diagonal towards the sea, but the smell of burnt wood hangs in the air as I stand by the river.

The woman moves slowly, like a heron walking, but she is not lame. White hair ruffles around her uncovered head and her grey cloak streams in the wind. When she gsets to the church she takes her hand from her cloak and throws something through the window-hole into the smoke. It is heavy like a pebble and glimmers. She turns towards the river and raises her arms, and then begins a low chant. Her voice is strong, and the words carried on the wind are Norse, not the Latin of medieval prayer. I think she knows I am here, but I cannot run from her and panic rises in my throat.

When I woke up the sun was stronger, and I was far too hot in my pyjamas and down bag. A 'two-person tent', I discovered, is big enough for one small person, some chocolate and a lot of books. I kicked one of the poles as I tried to wriggle out of my sleeping bag and the whole thing tilted sideways. I heard Yianni laugh outside.

'Don't stand there laughing at me,' I said. 'Hold it up while I get out.'

His shadow moved up the shiny pink canvas and he grabbed the apex of the poles. I tipped forward and stuck my head out.

I was glad I'd come. The valley is flat and green and the river runs like a road down to the rocky shore. The ruined farm buildings we've come for are scattered up the valley, mainly under the steep, scree-covered slopes that overhang the river and the bright pastures. The sea was dark blue that day and big pieces of bright white ice still drifted in the lapping waves, but the sun was warm and strong on my face and even Yianni, who if you remember wore a jumper in York in June, was wearing the shorts and T-shirt he had last year in Crete. I scrambled out into the sun and stood barefoot on the coarse grass.

'Sleep OK?' he asked. He's grown his hair, but the resemblance to that statue of Paris in the Cast Gallery is undermined by the beginnings of a beard.

'No,' I said. 'What's under that stone?'

'I don't know.' He patted my arm. 'I haven't lifted it up yet.'

'Then how do you know it's not a grave?'

'I don't. But it would be an odd place for it.'

'It's the odd ones you've got to worry about.'

'Nina, anyone buried here has been in the ground at least five hundred years.'

I looked round. Sun and purple pyjamas make wraiths seem unlikely.

'Any chance of some hot water for a wash?'

'No,' he said. 'There's a perfectly good river over there. We can't waste paraffin on washing, we need to boil the drinking water. Go on, it's invigorating.'

I looked at him. I think invigorating is what I said when we got him into the sea at Brighton. Nevertheless, cold is transient but dirt gets worse. I found my towel and headed for the river, picking my way over the prickly turf and thinking that the Norse women must have walked this way dozens of times a day. He'd put up four tents in the little field bounded by broken drystone walls, but there was no sign of any of the others. Yianni was sitting on a stone facing up the valley, writing in a notebook. I stepped across the pebbles on the river bank and dipped my foot in the water. Even though I knew it was meltwater from the glacier, the water was colder than you'd think it could be without setting. I looked up at the ice and clenched my teeth, knowing that if I didn't wash in the river I wouldn't be washing at all, not for weeks. Foot back on the warm stone, I glanced round at Yianni. He was still looking away, so I dropped my pyjamas on the rocks and floundered in, sitting down before my mind registered the pain in my feet and legs. For a moment I thought I'd never move again and would be found by the next generation of archaeologists, a mad Englishwoman frozen in a Greenlandic river, and then I crawled out, feeling the sun and breeze stroke my bare skin, and struggled back into my sun-baked pyjamas without drying myself.

Yianni looked up as I went by, my feet now numb to the prickly grass.

'Invigorated now?' he asked.

If I'd relaxed my jaw he'd have seen my teeth chattering, so I tried to sweep on as if I wasn't wearing wet pyjamas. There's no way of entering a tent with dignity, and it collapsed on me as soon as I crawled through the flap.

I heard voices while I was struggling to get dressed in a tent which kept subsiding, so I didn't hurry. I guessed the others had arrived but I could see no reason to expend energy trying to deal with whoever had escorted them, and I was hoping the horses might have gone by the time I came out. The horses were one of the things that could easily have stopped me coming. Yianni's always claimed that there's no recorded case of horses noticing and exploiting their superior strength and size, so I don't know why he let me come on the plane with the tools. He made the others travel by boat as far as the supply ship goes. After that there aren't any roads, it's boats, planes or horses. But he's promised he'll splash out so we can all leave with the finds on the little plane. When he said that I knew he was expecting burials to dig up, biological material that is, like me, too unstable for long journeys.

Anyway, the little plane from Nuuk was fun. There was a pilot called Anders with shiny muscles, and I sat behind him on a box full of trowels and divided my attention between him easing big levers up and down and the landscape, which was spectacular and interestingly close. It's hard to estimate distances over snow, but I realised quite how low we were flying when he pointed out some kayaks crossing an inlet like darting fish. It's not entirely reassuring to be told that it's safer to skim the ice than to take such a small plane through low cloud, so I imagined a future in which you became an Arctic pilot and we lived in a little house by the sea, with white walls and one of those cast-iron stoves. Which is not to suggest that I will ever consider changing my mind about living within walking distance of purveyors of fairly traded coffee beans and hardback books.

I could hear Americans demonstrating team-building skills and putting up tents with unnecessary self-assurance while I wrestled with my underwear. Even once I'd managed to put my bra on while more or less lying down, I didn't hurry. I had the feeling that four weeks with the people I could hear out there would be plenty and there was no sense in starting it any earlier than I had to. When I did come out, hair still a mess because there is no way of brushing long hair in a small tent, two men were leading an alarming number of horses across the bright grass down by the shore, and the field looked as if someone had cut-and-pasted microscope slides of fungus or bacteria onto an Arctic summer landscape. Round and oval tents in unnatural colours had spread across the turf, and the huddle of people on the stones by the river where Yianni had set up the stove looked as if they'd been imported from another image, probably the alumni magazine of some rich American college. I thought I would rather get back in the river than meet a bunch of confident strangers who would have to live with me for the rest of the summer, but I walked towards them, telling myself to listen when they told me their names. I didn't, you know, when I met you at Charles's party. When I went down the next morning and told Helen and Claire that there was a man still asleep in my room, they asked who it was and I had to admit that I had no idea. And then I didn't want you to think I was the kind of woman who slept with men whose names she didn't know so I couldn't ask. I looked at your post and would have called you Stephen had you not taken a message for him before I'd quite decided to risk it.

I was still thinking about that when Yianni introduced them so of course I forgot all the names immediately. There were several very clean-looking Americans who could rise from the rocks while holding cups of instant coffee (I thought Americans knew better than to drink instant coffee) and extending large flat hands and open smiling countenances like something out of Thornton Wilder. They were mostly wearing white T-shirts which appeared to have been ironed and were probably going to go on looking like that no matter how much mud and river water came their way. There was also a Scottish girl who appeared potentially congenial except that she exuded peaceful self-confidence, which is doubtless a fine quality to have but unnerving for the rest of us. I know you say people sense unease and become uneasy, a bit like dogs sensing fear and becoming aggressive. All it means is that the bad behaviour of people and dogs is my fault. I don't know how that's meant to help.

Everyone stood about as if the rocks had got too hot to sit on after we'd all expressed the statutory and fictional pleasure in first encounters. Reading Henry James, you'd think it's the Old World that's meant to be courteous but Americans practise levels of politeness unknown to the English bourgeoisie. The prospect of trying to beat American good manners before breakfast made me feel like a bird in a net. I went and sat on the grass next to the Scottish girl, who looked at the Americans and then at me and sat down too. Pebbles wavered through the clear water at our feet and a white cloud processed across the dark screes above the valley. Yianni turned back to the Primus and began to dole stewed dried fruit from a steaming pan into chipped enamel dishes. I could see that it was filling and would prevent scurvy. The Scottish girl was also watching him and it became clear that one of us needed to say something. I tried to relax my shoulders.

'Do you think the water is as warm as it looks?' she asked.

I glanced at her. She was looking at the river with a slight frown, as if she'd lost something in it.

'No,' I said. 'I tried to wash in it earlier. It feels colder than water.'

'Oh,' she said. 'Oh well. It's good clean dirt, on digs.'

'It'll have to be.' I couldn't imagine I'd be able to con myself back into the river now I knew what it was like. 'We can't put any soap or shampoo into the water anyway.'

'No,' she said. 'No rubbish, no fires, no soap. No picking anything that grows and no planting anything that doesn't. Good thing the Greenlanders didn't think like that, isn't it, there'd be nothing here for us to find.'

'Maybe they did,' I said. 'Maybe that's what happened. They weren't raped and pillaged by pirates or starved by climate change and they didn't all go to America or back to Iceland, they just went green and trod so lightly on the earth that nobody knows they were there.'

'Maybe,' said Yianni, passing us each a bowl. 'But there'd still be bodies. Even if they had wicker coffins. Or ashes. You can't just disappear the dead. That's the point about archaeology. People can't help leaving themselves.'

I shivered. The enamel dish was too hot and I put it down.

'I'm sorry,' I said. 'I've already forgotten your name. I'm awful at listening to introductions.'

'I'm Catriona,' she said. 'And you're Nina? Are you the one from Oxford, with the scholarship? Yianni told me. You must be really good.'

I have no idea what the right answer to this might be. Yes, I'm brilliant. No, I'm very thick but good at deceiving learned committees. The cloud had nearly passed the mountain.

'I live in London now,' I said. 'My partner has a job there.'

I wondered if you had finished painting the bedroom and how my orchid was responding to your ministrations. I thought of you standing on the step locking the door on a silent flat each morning and returning to the takeaway menus on the mat and the crumbs on the table in the evening.

'Do you like it?' she asked.

'No,' I said. 'I miss Oxford. But David likes his job, and I can work just as well in the British Library as the Bodleian. I just happen to prefer old buildings where my friends are to curved wood and strangers.'

'But you do the nineteenth century, is that right?'

I wondered why Yianni had told her about me and not me about her, and what else he might have said. Do you warn people, before they meet me? I picked up the dish again and prodded a grainy pear.

'Yeah,' I said. 'English lit. I'm looking at the influence of Old Norse sagas on Victorian poetry. Mostly the Pre-Raphaelites so far, though I'm getting into ghost stories. And then I got a grant. Well, not really. But there's a bequest fund you can apply to for research-related travel. I just need to write something explaining how being here helps with my doctorate. It doesn't, really, that's the whole point, that the Vikings turned into a Victorian fantasy, but I'll make something up. I've always wanted to go to Greenland and Yianni said I could come if I didn't mind being unskilled labour.'

And when he said 'unskilled labour' you put your glass down and said everyone should try manual work once in a lifetime, you in your hand-made shirt, and when we laughed you said you'd helped in the conservation department when you started at Sotheby's. At least I get soil under my nails. If not worse.

'You do medieval archaeology?' I asked.

She nodded. A breeze stirred her Flemish Madonna hair.

'Faroese, mostly. Early medieval North Atlantic migration patterns. I end up reading bits of oceanography as well.'

'Has the Atlantic changed in eight centuries?' I asked.

'Well, some people think so and some not. Temperatures change but no one's sure what effect that has.'

One of the Americans leant forward, a short guy with red hair that stood up like thistledown and an All-American jaw-line.

'Cool thesis,' he said. 'Funny how other people's doctorates are always cooler than mine.'

'You're not American,' I heard myself say.

He looked at me. 'Should I be?'

The blades of grass are hard to tear, tougher than the green stuff at home.

'No. I just thought you all were. Sorry.'

'What's your doctorate?' asked Catriona, eating an apricot that lay in her spoon like an egg yolk.

'Cultivation and foraging in liminal settlement areas in Norway,' he said. 'I'm working with Brian Claridge, at Madison.' He looked at me. 'So you're right, I am based in the US. I'm from Sheffield.'

'Sorry,' I muttered. You tell me to pretend to be confident. I took a breath. 'What are liminal settlement areas?'

'Places where people can only live in good years,' said Catriona. 'Right on the edge of habitability.'

The red-head tipped his bowl. 'So in northern Norway, you find houses or even villages that seem to have been deserted for a few decades then rebuilt and then deserted again. The main stress factor is plague but short-term variations in climate do it as well.'

I poured juice from my spoon back into the bowl. I could see why he thought other people's research was more fun. 'You could use that to analyse the property market.'

'I expect someone does,' said Catriona. 'I met Brian Claridge at the NACR conference last year. Were you there?'

The deceptive All-American jaw was chewing but he shook his head.

The other guy put down his empty plate. He was tall, with those big American shoulders that bespeak a childhood diet of beef full of growth hormones. 'I've got a friend who's working on the anthropology of surfing at the University of Hawaii,' he said. He really was American.

We all looked out at the ice gliding across the black water and the river swirling over the pebbles and saw the point of Hawaii.

'My friend Mike wants one called "Consuming Passions:

Restaurants in Twentieth-Century French Film",' I offered. 'But what he actually does is auditory neurophysics.'

'Sounds more lucrative than French film,' muttered Yianni. 'Or medieval archaeology.'

'I'd like to be in Venice.' Catriona put her bowl on the rock and stretched out her legs. 'How about, "Representations of Power in Seventeenth-Century Venetian Portraiture". Though I always wonder about the stuff that's never been called up in copyright libraries. You could do something on the unread holdings of the British Library.'

'I know,' I said. 'I always think the most interesting work would be on the bits of history that got lost. I mean, that's a lot of the appeal of the Greenlanders, isn't it? We want everyone to leave a story. And those lost Americans. You know, the early colonists. The ones who just disappeared.'

'Roanoke,' said the fake American.

'Anyway, I bet the unread holdings of the BL are probably mostly railway handbooks and things,' I said. There were apple rings left cold and wet on my plate and I thought they could probably stay there. 'That and things people would be too embarrassed to read in Humanities Two. Mills and Boon. Venice sounds better.'

The American woman cleared her throat. Her hair was perfectly tidy, as if she was expecting to be photographed, and I saw that she was wearing make-up, the kind of expensive, cunning make-up that betokens years of practice. It looked as if someone had dropped a Barbie doll on the grass. I found myself fingering a spot on my chin that I'd earlier decided didn't exist as long as I didn't have a mirror to see it. I thought

The Beauty Myth was compulsory reading for preppie American women, so often in search of victim status.

'"Reading and Domesticity in Nineteenth-Century American Children's Literature,"' she said. She didn't sound as American as she looked, one of those voices drifting in the mid-Atlantic that sounds fake whichever side you're on. Her stewed fruit was still floating like dead goldfish in her bowl. You'd have to be seriously screwed up about food to worry about the calories in a stewed apricot.

'What in particular?' I asked. 'I do nineteenth-century English lit.'

She addressed the ground at my feet.

'Oh, I've always liked *Little Women*,' she said. 'And Laura Ingalls Wilder.'

'Hoop skirts and home baking,' I said. 'There were women in nineteenth-century public life, you know. Qualifying as doctors and campaigning in politics.'

She moved her gaze to my jeans. 'I know. But there were hoop skirts and home baking as well.'

Yianni stood up.

'If you've all finished breakfast, let's look around the site. I'll wash up while you get your notebooks.'

'Where do I brush my teeth?' asked the woman with the hair.

Yianni grinned. 'Anywhere you like. There's drinking water in the stores tent. But don't spit the toothpaste.'

'What am I supposed to do with it?' She seemed to be asking a rock to the left of his shoulder.

He shrugged. 'Swallow?'

The fake American smirked.

'That's what I did last night,' I said. 'It can't be poisonous.'

She took a breath and then shrugged. 'OK.'

'I'm sure we'll get used to it,' said the Hawaiian surfer's friend. 'Just use less.' We walked back towards the camp in silence and I watched as the others crawled into their tents. Catriona and then the tall guy came out with spiral-bound notebooks. I'd brought this notebook and one for 'research', whatever form that might take, so I found the back of the printout of my flight times that you made while I showered that last morning at home. It occurred to me that pieces of paper would not rise from the turf the way they rise from all flat surfaces at home. I watched a sheep wander between the tents, cropping industriously. The components of vellum are still more readily available here than paper.

Yianni was standing at the other side of the river, among sticks and strings marking out a grid across crumbling stone walls. Even I could see ridges in the turf and oddly square patches of vegetation, and at one side there were lines of stones which I couldn't imagine had really been there for eight hundred years. A shallow trench still ran from the river through the fallen stones and down to the sea. The medieval Greenlanders had running water, did you know that? And saunas and frozen-food stores in the cellars.

Catriona joined me.

'Has he already shown you round?' she asked.

I shook my head. 'I only arrived last night. There were all the stores and tools to deal with. I probably know less about the site than you do.'

'I only know what Yianni's told me,' she said. 'It's an odd one, easily accessible from the sea but probably inhabited late. Mostly the late ones are inland. The ones you can see from the open sea got raided as soon as the cod fishermen found them undefended.'

We set off across the field.

'Is that what you think finished them off?' I asked. 'Pirate raids?'

Out at sea, the horizon was a straight line where dark water met grey sky. You'd see anyone coming long before they landed. But maybe not before they saw you.

'Some of them,' she said. 'And some of them probably found their farming methods weren't working anymore when the climate cooled. I don't think there's a dramatic end, just life getting steadily more difficult for a few generations, maybe raised mortality and less food, until the people who could leave left. But most of the sites here look more like the Clearances and Ben's liminal settlement areas than the Potato Famine. There'd be more burials, and mass graves, if they had plague or acute famine.'

I thought about the headlines again. Mass graves. We were coming to the river, and the Americans were already finding stepping stones.

'Did you hear any news, on your way here?' I asked.

'Not especially.' She put her notebook in the pocket of her green cagoule. 'Oh, you mean the virus thing?'

'Mm.' A fish flicked the surface of the river.

'It's just a media panic. I wouldn't worry. Easy journalism for August. Remember last time, people were actually buying masks and spending God knows how much on fake vaccine on the internet and then the papers lost interest and we got scared about something else. Honestly, by the time we get back everyone'll be worrying about the property market again.'

'My partner David says it's a smokescreen for something else. Either the Americans are going to say terrorists have been spreading germs so we need to invade somewhere else with oil or they have invaded somewhere else with oil but Americans are too scared of other people's handkerchiefs to notice.'

'He likes his conspiracy theories, then?'

I looked up. 'There was that thing about crop sprayers.'

'Precious little evidence for it.'

Our own Americans were reaching the other side of the river, having repositioned a series of rocks so the woman could avoid getting her feet wet. The tall guy gave her his hand as she made the last jump.

'Anyway, a proper pandemic might be quite good for the environment,' said Catriona. 'It's probably about the only way of arresting climate change now. Depopulation from the plague did wonders for medieval fauna and flora. But last I heard it was a few children in Delhi, a hypochondriac American vet with a cold and maybe some wild birds. Are you going first or shall I?'

'You,' I said.

She stepped across the stones as if they were a zebra crossing. I followed quickly so she couldn't see my alarms and hesitations, making the last leap by concentrating on the skyline where the rocks rose from the slopes of scree. It wasn't cold water I feared, but humiliation.

'So this is the big farm,' Yianni said. The walls were waist-high and several feet thick. 'We're standing in the hall. Those are the ante-rooms.' He pointed to more heaps of stone.

'Is that the lintel?' asked the tall guy.

'Yeah,' said Yianni. He pointed. 'There's the mantel. It was some fireplace.'

There was a long, flat rock, as big as the standing stones we used to walk round in the Dales when my grandparents were alive.

'How did they lift that?' I asked.

'We think they had rollers,' said Yianni. 'There are bigger stones than that in the church. The trench, of course, brings water.'

'They had a good view,' said Catriona, standing by the lintel-stone. 'Odd, though, to face the door into the prevailing wind. Or do you think this has been moved?'

'Don't think so,' said Yianni. 'Maybe it was worth the wind to be able to see what was coming.'

'There's a cross on the lintel,' said the smaller guy.

'I know,' said Yianni. 'And it's old. But of course it could have been carved in situ at any time.'

I stood next to Catriona. The view of the hill and the river curving down to the stony beach was exactly what I'd been imagining at my desk at home, and I caught myself thinking the site was wasted as a ruin and wanting to do it up and move in. The opposite of archaeology.

'I told you in the briefing papers that this is the site Norman MacDonald identified with the farm owned by Bjorn Bardarson in *Bjornsaga*. Late thirteenth century. The saga says his brother burnt down the byre. This byre was burnt and doesn't seem to have been rebuilt, and once we get the dating done we should know when the farm was abandoned. We've got farm V49 fifteen kilometres south, and I spoke to Adam Morris about his work there before I left. He's sure it was still occupied in the late fourteenth, so if this was abandoned earlier we probably can tie it into the saga. But we'll see.'

He started talking about the lab analyses of V49 and Adam Morris's unpublished work. It was still warm, but there were grey clouds gathering in the northwest and the sea looked duller than it had. *Bjornsaga* is the one with Ingibjorg and Kristin. They were twin sisters, and their father was tangentially involved in a long-running feud of the sort that defines most of the sagas. One day Kristin was found dead and 'unpleasantly damaged' on the beach. No one knew who'd killed her, or trusted their guesswork enough to attempt vengeance, which was the usual way of stopping the dead coming back. They buried her quickly and thoroughly because everyone expected someone killed like that to make trouble. She did. Every night she came creeping into Ingibjorg's bed in a state of advancing decomposition, muttering allegations, until Ingibjorg 'spoke no more sense but uttered strange prophecies until she died.' After that one or other of them often sat on the roof of the house and woke people by shouting, but that bothers me less. It's the idea of someone who loves you turning into a revenant who comes to decompose in your bed and drive you mad that's particularly disturbing. Would you rather be haunted by your rotting beloved or lose her entirely? I think I'd be good at haunting.

'Any questions?' Yianni was saying. 'Yes, Ruth?'

Ruth, still talking to stones and thin air.

'Is there any material evidence of late external contact from V49?'

Late external contact, I think, means interaction with the outside world in the final decades of the settlement.

'Nothing conclusive that Morris was prepared to tell me,' said Yianni. 'But he wouldn't let me see all of his data.'

'Crazy,' said the smaller guy. 'You could hardly publish his data while you're in Greenland.'

'Well, it's his prerogative,' said Yianni. 'Come on, there's more up the hill. And of course the church.'

I hadn't noticed the building up the hill before. It was partly sheltered by a little knoll and much more complete than the farmhouse, the walls still shoulder-high and the stone door-frame still standing.

'Barn,' said Yianni. 'With ante-rooms.'

He pointed to rubble lying in straight lines across one side.

'In use later than the house or just better protected?' asked Catriona.

'Bit of both, maybe,' he said.

It's nice, the barn. You could try a garden, at least some flowers on the sheltered side, though it would have to be something that didn't need much sun. Cathedral ceiling, solid fuel stoves. No gas or electricity, of course, but this is a place where a domestic turbine really would earn its keep. There was an article in the interiors section of that last *Observer*, which I left on a circular bench under a palm tree in Copenhagen airport, about barn conversions. A derelict barn costs more than a house now in some parts of the Home Counties. I have no idea about Greenlandic property prices but I bet if we sold the flat . . . I caught Yianni's eye and blushed as if I'd been caught thinking about sex. The others were writing things in their notebooks.

'So we'll make a start after lunch,' he said. 'There's something I want to say right at the beginning: we've got less than six weeks here, and if we get persistent rain we may be held up, so let's make the most of our time, OK? You know how fast the season changes here and summer's half gone with marking exams before we even start. We are going to be starting early and finishing late and I'm not planning days off.'

'Sure,' said Catriona. 'It's not as if we'll miss out on the vibrant nightlife of rural West Greenland.'

'Fine by me,' said Ruth.

I thought about my books.

'My grant requires me to do some writing,' I said. 'It's not just data, I can't write it all up when I get home.'

'You'll have time to write, Nina. It's light nearly all night anyway at the moment.'

Ruth looked at my feet and pursed her lips as if my boots were too obviously last season. 'We've all got work to do, you know. You're not the only one.'

I didn't see why I should have to work in the middle of the night but I thought I might find it easier to persuade Yianni on my own later.

Lunch was water biscuits and cream cheese rendered no less nasty by alleged smoked salmon flavouring, followed by powdery red apples.

'This is all the fresh fruit,' said Yianni. 'When these are gone, it's dried fruit and vitamin supplements.'

'Then why on earth did you bring tasteless American apples when the English season is just beginning?' I asked. 'You were flying stuff in anyway, you could have added some decent fruit. There are heritage varieties in all the farmers' markets at the moment.'

'Oh, for pete's sake,' muttered Ruth. I looked at her. 'What?' She was still focusing on my boots. 'Nothing. Never mind.'

'They're just apples, Nina. And there are some lemons.' Yianni picked one up and turned it, watching the sun on its unnaturally shiny surface.

'To ward off scurvy,' suggested Catriona.

'And a ration of rum?' asked the fake American, who turned out to be called Ben.

'Bad luck,' said Yianni. 'But there are onions. Otherwise it's all dried or vacuum packed.'

'Bet the Greenlanders did better than that,' I said. 'Didn't they have mussels? And cloudberries?'

I am not sure exactly what a cloudberry is but it sounds much nicer than dried figs.

'They were farming and fishing,' said Ruth, spreading pink paste on a cracker. 'We're digging. Didn't he warn you? If you wanted a gourmet dig you should have written about the Romans.'

'Speaking of which,' said Yianni as he stood up. 'Catriona, could you tidy up here? Let's get going.'

And then we spent all afternoon digging. Apparently the interesting bit comes later, and in places where there are roads archaeologists sometimes use a mechanical digger for the first stage. But the sun shone, and moths fluttered out of the heather, and the turf smelt like clean laundry. There were flowers like overgrown buttercups, and the sheep, which stared at us with alarming malevolence at close range, kept a polite distance and provided a bucolic soundtrack. As manual labour goes, it was the kind of thing rich people pay to do on holiday and, with the odd pause for water and polite exchanges, we kept doing it until late afternoon, when I began to realise that digging requires muscles undeveloped by reading, typing and the occasional yoga class. I tried a prayer stretch on the grass.

'Stiff?' asked Catriona.

'I will be. Aren't you?'

'I've been changing hands. But yes.'

She put down her spade and stretched her arms over her head.

'Taking a break?' asked Ben, standing up and circling his shoulders.

'It's past tea time,' I said.

'You know Yianni better than we do.' Catriona turned her head from side to side. 'But he didn't strike me as a great proponent of the tea-break.'

'No,' I admitted. 'Especially not here. He probably thinks it's a waste of paraffin.'

'Don't tell me you put paraffin in tea,' said Ben. He looked up like your parents' dog waiting for a biscuit.

'Not in Scotland,' said Catriona. She sat cross-legged and I watched a moth crawl up her trouser leg. The grey clouds were swelling out at sea but our shadows were still sharp on the turf. Ben paused.

'Hey, that's an Arctic Blue. I've never seen one before.'

'A what?' I looked at the sea, as if an Arctic Blue might be some kind of iceberg.

'Butterfly. Look.'

It was 'blue' in the sense that those grey cats are blue, as if people have to justify their taxonomic impulses by making things sound more surprising than they are.

'You know your butterflies, then?' asked Catriona, watching as it danced away.

He shrugged. 'A bit. I like to get out onto the hills. It's Louise, really, my girlfriend, who's into moths and butterflies.'

'An entomologist?' I asked.

He was scanning for more Arctic Blues. 'Geography teacher.'

I suppose a PE teacher would be worse. I looked at Catriona but she was watching the sea.

'Sheffield or Madison?' I asked.

'Sheffield.'

'How's that work?'

He looked as if I'd asked what kind of contraception they favour.

'Fine. Why?'

Catriona stretched. 'I suppose we should do some more work.'

Jim came over and sat beside her. 'I'm still stiff from those horses,' he said. 'Though it was a great way to arrive. I haven't ridden for years.'

'You used to ride regularly?' asked Catriona.

He circled his head. 'When my grandpa was alive. They had a farm.'

'Did the Greenlanders have horses?' I asked. Icelanders do in the sagas, but I couldn't see why they'd be useful in a place where small boats are the main means of transport.

'Some of them,' said Ruth, from behind me. She came and stood over us. 'Rich farms would have had a pony or two. Icelandic ponies.'

I craned round and found myself looking up her nose. I shuffled forwards. 'Are you working on Greenland?'

'Not really. First contacts in North America. I've got a chapter on the Vinland sites.'

'But you fancied Greenland?' I asked.

'It seemed a good chance. To get away.'

I wanted to ask her what she was getting away from.

'Was your supervisor OK with that?' asked Catriona.

Ruth still didn't sit down. It was hard to see her face.

'Oh, yes,' she said. 'He's very flexible.'

That got us onto the Supervisor Conversation. You'll be pleased to hear that the British are still much more colourful than the Americans, who seem to have a 'meetings schedule' set by the department as well as all those classes, and of course don't have the same drinking culture. I started telling the Verity Buchan story, and had just got to the bit where she was lying on a sofa in her knickers pretending the teddy bear was talking about Piers Plowman when Yianni came over. He can't have heard it before – do you think? Anyway, he waited until it was over but then said there was time for more digging and he was hoping we'd have all the turf up by bedtime. Which made it sound as if he was thinking of working days in terms of tasks rather than time, and also as if he was prepared to make the best of the light nights. We went back to work.

I thought I'd sleep well, after that. I moved my tent off the cannibal-monster-infested tombstone and onto some well-sprung turf next to Catriona. Everyone went to bed soon after dinner (noodles and bottled sauce – it's not scurvy but rickets we need to worry about here). I've brought *I Capture the Castle* in case of need and I read a bit of that and settled down. I slept fast enough, but had bad dreams.

The old woman is riding up the valley bareback on a horse too big for her. I run easily alongside, breathing smoothly, as if there's no weight of blood and bone. As we come over the hill I can see another farm, and everyone is still there. Two men are doing something with fishing nets in a boat pulled up on the beach, and a man is digging the infield. There's a woman sitting on the stone bench by the door, nursing a bundle in a woven shawl that murmurs at the breast. Everyone looks up as the horse picks its way down the stony hill, but the men keep working. The woman stands up and moves awkwardly towards us, the baby held in one arm and still sucking. She knows what the old woman is going to tell her. She has seen the smoke rising from the next valley, and the men told her how they hid in the little boat behind an island and watched as the fishermen's ship scudded away. They have not told her about the cries of the women, or about what they found at the other farm. Not about the little boy, or the puppy skewered at his feet.

Something woke me, a sound. I sat up, waiting for the drumming in my ears to slow so I could hear the night. Dreams are nothing new, but I don't like it when I wake up and I'm still there. It was light, but dim, and I was afraid to open the tent and look at the cold sun. The sheep and birds were quiet, and after a few minutes I thought I could hear ragged breathing and sniffing, like someone quietly crying. It was very near.

'Who's there?' I called.

The noise stopped, but no one answered. I lay down, scared to sleep again, and thought of you.

The baby is crawling on the worn grass at my feet and a little boy is sitting nearby, keeping an eye on the baby but also carving a small piece of wood. I follow the baby's mother down to the river, which gleams through the grey fields in the fading light. I stand behind her as she lifts the heavy wet wool from the pool where it has been soaking, and the water runs down her arms and splashes her brown dress. My foot slips on a stone and she turns with a gasp, the wool clasped in her arms. She faces me, looks around, and shivers. She turns to the sea as if looking for storm clouds, or worse. The sun reaches the bank of dark cloud over the sea to the west and the air is suddenly colder. I am still, still and cold, and I want only to come back home.

And then yesterday we got up, ate dried fruit, dug all morning, ate crispbread and that extraordinary Norwegian cheese that tastes like a mixture of rust and condensed milk with the last of the apples, dug all afternoon, ate pasta with Textured Vegetable Protein (the texture only makes things worse) and tomato puree out of a tube followed by chocolate, and went to bed. I began to suspect that the practice of archaeology is less interesting than I'd hoped. All that came to light were worms, which are why I don't like gardening, and Yianni spent the whole day generating paperwork. In the night I heard crying again.

There are ships on the horizon. The woman and the baby stay in the house, and she pours water onto the fire. It runs across the floor, black with ash, and marks the edges of her dress like blood. Without the fire it is dark inside. The baby cries and the mother picks it up, but her eyes are fixed on the door as she offers it her finger to suck. It slurps furiously and cries again. The boy is huddled in the sleeping place. I open the door and slip out.

It is bright, out here. One man is down on the beach, throwing seaweed over the little boat that has been dragged across the pebbles to rest behind a rock. The boat may not be visible from the sea but the furrows in the stones are obvious. The other two are gathering the sheep into the barn, keeping low to the ground. The racks of drying fish stand like beacons on the shoulder of the hill. Inside the house is a stupid place to hide, and I make for the big rocks which dot the turf above the barn.

The zip on my tent ripped open and I sat up and screamed. The shape pushing its way in raised its head. Yianni.

'Nina, hush! It's me, it's all right.'

'What in hell's name are you doing here? Jesus Christ, it's the middle of the night.'

'Nina, you were calling out. You've been muttering for hours. It woke me.'

'I wasn't,' I said. 'Yianni, it wasn't me. I hear things in the night too. I don't like it here.'

He reached out and patted my shoulder.

'It's bad dreams, Nina. Go back to sleep. It's all OK. Look, just sheep and sky out there.'

He held the flap open. It was not dark but the sky looked

dead. The sea was black and quiet, and there were no birds.

'It's weird,' I said. 'What time is it?'

'Just gone one. The sun will be back in a few minutes. You can still get a good sleep before morning.'

I woke to voices and the purring of the Primus. The tent was hot and I could smell apple crumble, although I knew it was only stewed fruit. You won't be making crumbles, will you? And I bet you've gone back to having cereal for breakfast, soggy in skimmed milk. I sometimes suspect you are interested in food only to please me, that you might be equally happy with someone who uses custard powder and generates less washing up, not to mention storms over curdled eggs. I put my jeans and jumper back on and crawled out. Jim and Yianni were sitting on the stones near the tents, the Primus resting between Yianni's feet like a family pet.

'Good morning, Nina. You didn't sleep too well?'

'I hope you weren't disturbed.'

Jim looked improbably clean and freshly shaven beside Yianni's stubble and grimy sweatshirt. Shaving seemed so unlikely, with icy water and no mirror, that I wondered if he used depilatory cream instead. He watched as I tried to push strands of greasy hair back into my plait. Or maybe Americans get their chins lasered before going abroad.

'Well, you can't help your dreams.'

'No,' I said. 'Sorry.'

Ben came across the field from the river, looking invigorated.

'Is it more of the same today?' I asked. 'Digging?'

Yianni pulled his sleeve down over his hand and lifted the lid of the pan at his feet. A cloud of apple-scented steam drifted between us.

'That is what we're here for,' he said.

'I know. I'm just wondering when we start finding things.'

'When we find them. That's archaeology.'

Jim ran a broad hand through his combed hair.

'What got you interested in Greenland, Nina?'

Behind me, I heard Catriona start rustling in her tent.

'I like the far North,' I said. 'I always have. Hence the thesis, really. I thought there might be some good travel grants, and I fancied a few months in Iceland. Though that was before I met my partner.'

'Less keen on travelling now?'

Not less keen. Only that the price of separation from you is too high to pay.

'He works,' I said. 'In a map and print gallery. He can't get six months off to go to Iceland and follow in the footsteps of William Morris. And I don't need to, really. In fact the imaginary nature of Iceland in Victorian poetry is the whole point of my thesis, but it would be fun. And there are grants.'

Catriona crawled out of her tent. She hadn't brushed her hair either.

'I thought Iceland was amazing,' she said. 'I wanted to paint everything. I'd like to go back with oils and great big canvases.'

'You paint?' asked Jim. As if he'd worked it out for himself.

'I'm better at painting than research,' she said, sitting down. 'I didn't go to art school because I thought I ought to do something more practical. And now I'm writing a doctoral thesis on medieval history.'

'Did you bring paints here?' I asked.

She looked at the sky and the sea. 'Yes. I've always wanted to paint proper ice. I sometimes wonder if I specialise in the North Atlantic because I like painting the light.'

'You won't get proper ice, you know,' said Yianni. 'At least, I hope not.'

'It's pretty good,' said Catriona. 'I'm not complaining.'

After breakfast, while Ben and Jim were down at the river rinsing plates and Ruth was in her tent, presumably enacting one of her beautifying rituals before facing the day's stones and sheep, Catriona went to her tent and came back with an artist's pad. It was my turn to make lunch and I had been considering ingredients and scanning the ground for wild herbs.

'Did the Greenlanders eat angelica?' I asked. 'It's all over the place. You can buy it candied, at home. Recipes tell you to use it to decorate trifle.'

'If it's edible I should think they ate it,' she said. 'Would you like to see my paintings? I thought you were interested.'

'I'd love to,' I said. Which I meant, except that even four years in your company has not taught me to say intelligent things about visual art. I cannot, really, get beyond what I like and dislike. Sorry.

But Catriona's paintings are beautiful. She paints in Skye, watercolours, mostly seascapes, and she makes the wateriness part of the interpretation, as if it's overflowed from what she's looking at onto the page. There are shapes in her seas that bulk and turn like seals, or drowning men.

'Do you exhibit?' I asked. 'These are fantastic. Have you done any here?'

'Not yet. I'm thinking about it. It's why I came, to be honest. Don't tell Yianni.'

'I won't. But he wouldn't mind. I came because it sounded fun. Because I thought I'd probably like remembering it later on.'

'But you're his friend. I'm meant to be here for work. I had a few paintings in an exhibition in Edinburgh last year. Just a local gallery. Nothing big.' The book in her hand shook a little.

'It sounds exciting to me,' I said. 'Did they sell?'

She nodded, biting her lip and smiling like a child remembering Christmas.

'So the gallery wants more?'

'Mm. Well, they want to look at more. They might not like them, of course. Those ones were less abstract.'

'These look lovely to me,' I said. 'I don't know much about it, but I do get stuck in galleries while my partner's networking. I get bored, usually, but I could look at these for a long time.'

'I'm glad you like them,' she said.

Ben and Jim were coming back, carrying the dishes in the wicker basket so uncharacteristically provided by Yianni for the purpose. It's like the one old Mrs Rabbit takes to buy five currant buns and a loaf of brown bread.

'I'll put them away,' she said. 'Don't tell people.'

I wasn't sure what I wasn't telling people.

'I won't,' I promised.

After lunch – crispbread and tinned tuna with mayonnaise from a jar – Jim asked Yianni if he could look at his e-mail. I'd been trying to postpone asking for as long as possible, knowing that Yianni would get cross about what he regards as recreational use of the connection, and knowing also that no amount of e-mail would be any replacement for your voice in my ear and your arms around me.

'Is there a special reason?' Yianni asked. 'Once a week was what I said in the briefing. The satellite connection's expensive, you know.'

'No.' Jim scraped at a smear of mayonnaise with his finger. 'I just worry about my family, is all. There are a lot of poultry farms where I'm from. I haven't picked up any news in a while.'

'What would you do?' asked Yianni.

'I don't know. Nothing, I suppose. It's just kind of strange, though. I guess we get used to 24/7 news. I'm an internet junkie, at home.'

'Me too,' said Ben. 'I always get a shock when I look at my history. How much time I spend on-line when I think I'm working. Makes you wonder, doesn't it, if people wrote their theses faster before the internet.'

'Shouldn't think so,' said Catriona. 'Imagine all that extra time tracking down information. And travelling to libraries. Seems only fair, the internet gives us time and we give it to the internet.'

'I threw away the network cable for my laptop,' said Ruth. She was breaking a piece of crispbread into smaller and smaller bits. 'I have to go up to campus for the internet. You can't do anything about the news, anyway.'

'You don't use the net at all?' asked Ben. 'You don't check your e-mail?'

She shrugged. The bits of crispbread were so small she had to crumble them between finger and thumb.

'Most days. I go in most days anyway. There's not much that can't wait a day. People could call me.'

Could. It didn't sound as if they did.

'Anyway,' said Yianni, 'I think we'll stick to once a week. And please don't spend long browsing, it really is unbelievably expensive. We can take turns if you like, allocate days, and then someone can have a look at the headlines a few times a week. Use it this evening if you like. But me, I'm enjoying the break. Ruth's right, there's nothing we can do about news. Concentrate on what's here. There should be enough discoveries up there. Speaking of which?'

We started to get up. Catriona stacked the melamine beakers we used for water, chlorine-scented by the 'purifying tablets' that are supposed to counteract whatever people catch from sheep shit, and Ruth gathered the plates. She had a French manicure, new or at least touched-up since we got here. You know, that thing with three layers of nail polish that your sister does.

Up at the farmhouse, we had at last finished lifting the turf. Before the ground dried out, you could see lines and shapes in the soil, and Ruth and Yianni arranged stakes and string around the colour changes. I expected a plan of something to emerge from the ground, but it didn't.

'Right,' said Yianni. 'Trowels, now.'

He handed round what looked like plasterer's trowels, and allocated sections of the main hall. I found myself between Jim and Catriona. Yianni showed me how to dig, as if scooping spoonfuls from the crumbly soil, and I settled to work, alternately kneeling and squatting by the fallen walls. The sun was out, and I looked away from a worm writhing in the roots of the torn-up turf. I hadn't been going long enough to get bored when my trowel scraped something hard.

'There's something here!' I said.

''S probably a potsherd,' said Catriona. 'I've got some, look.'

She had a small handful of flat brown things which she was collecting on a stone. I brushed the earth away from where I was digging with my fingers. She watched and laughed.

'You get used to them,' she said. 'I'd forgotten. The first ones are exciting.'

I picked it up, lighter and flatter than stone, a rough rectangle about an inch across.

'Earthenware,' said Catriona. 'Here, put it in a bag.'

There are little plastic Ziploc bags of the sort used for beads or cheap earrings at home, but instead of a price you write where you found the thing in relation to the grid of string, the date and the finder's initials. You could almost reconstruct the site, later, and I suppose that's what they do in museums.

'Is this Norse, then? The last person to touch it was one of the Norse Greenlanders?' I asked, touching it through the bag.

'Probably,' said Catriona. She kept on digging. 'I mean, it's in the right place. As far as we know, no one else has been round here dropping plates. The Inuit didn't fire pots. Of course, it could be later, though in a way that would be more interesting since we don't think there's been anyone later. We'll find hundreds of them, honestly. By the end of today you'll be putting them back because you can't be bothered to go get a bag.'

Jim stopped and looked up. 'I hope not.'

'No, OK, not here. I've been on digs where people did. Do you think we'll find willow pattern here?'

'Willow pattern?'

My grandparents had willow pattern, including a cake-stand which my grandmother used to cover with doilies and home-made meringues, scones and fruit cake. She taught me about turning the oven off and leaving the meringues overnight.

'It's a running joke. Wherever you are, whatever the site, there'll be willow pattern. You'd think Wedgewood paid people to drop fragments into Roman remains. I think you've got to go back to the Bronze Age to escape willow pattern.'

'Or cross the Atlantic,' said Jim. 'But I've heard about it.'

My trowel struck something else and I pulled out another potsherd, darker and thicker.

'It's not from the same thing,' I said.

'They almost never are,' said Catriona. 'If three match, start paying attention.'

She added another one to her pile.

'How do they get here?' I asked. 'I mean, does it mean the Norse Greenlanders smashed a lot of plates? And left them on the floor?'

'Shouldn't think so,' said Catriona. 'More likely, they just had an awful lot of pots. There wouldn't have been many other kinds of container. And if this house was inhabited for, say, three hundred years, you'd expect a lot of bits and pieces.'

I hadn't thought about that, the idea that the farms were really old when they were abandoned. I had imagined people bailing out of a new and insecure situation, but most of the settlements were at least the equivalent of Georgian. As old as your parents' house.

'Why do you think they left? If they did?' I asked. 'I know nobody knows, but what do you think?' I couldn't help glancing out to sea, checking the horizon.

Catriona went on burrowing with her little trowel.

'Probably a bit of everything,' she said. The roots under her fingers were dusted with a crumbly loam that looked like coffee grounds. If Yianni serves up any more weak instant coffee I might be tempted to try them. 'Some of the farms must have been marginal even in good years, and when temperatures fell they'd have been unsustainable. And I think there is evidence for the erosion theory. That the farming techniques they were using exhausted the land after a while. There must have been some raids by fishermen once the cod banks were discovered, there are a good few burnt buildings. And when that happened in the west of Ireland they tended to take all the young people as forced labour. We don't really know how many Greenlanders went to America or what happened to them, and we've no idea how many went back to Iceland. It's not a dramatic story, my version, but I can imagine that over a century or two lots of young people might have left and the older ones have done less and less on the land. It's happened in parts of the Highlands and Islands even in the last century.'

'So you think it was one of Ben's liminal settlements?' I started to dig again.

'Kind of. But politically as well as ecologically liminal. They needed amicable contact with the rest of the Norse world and they had no way of defending themselves against the fishermen.'

Jim was working through his patch as if winter were on the way and he was planning to hibernate in it, but he paused for a moment.

'It's not the PC version, but I bet there were clashes with the Inuit. When the mini Ice Age came and the Inuit came down here. Two sets of people competing for the same resources during the winter, there's no way there wasn't conflict. And think about the way the sagas write about the Inuit, it's obvious they thought they were savages. Barely human.'

'Maybe,' said Catriona. She pulled out another piece of earth and brushed it off, then dropped it. 'Just a stone, that one. But the odd scuffle wouldn't finish off all these scattered farms.'

'And there was the Black Death,' added Jim. He was frowning at his little earthworks. 'That thing they say, anyone who left Iceland with it would have been dead before they got to Greenland. My supervisor says, the thing is, you don't have to be alive to give someone the plague.'

Or any other kind of virus. My pulse quickened. Catriona pushed her hair behind her ears.

'Yes,' she said. She put her trowel down and started to arrange her potsherds against the handle, nudging each one with her forefinger until they were perfectly aligned. 'I don't think this is the time to think about that, is it, but the burial patterns right along the coast are wrong. Epidemics leave mass graves.'

Mass graves again. I realised I was holding my breath and tried to exhale.

'You wouldn't get mass graves with all these isolated farms,' said Jim. 'And there are a few where whole families have been found in the beds or around the house.'

My hand shook. 'Can we stop talking about epidemics, please? I'm going to the loo.'

Jim shrugged. 'OK. If it bothers you.'

I put my trowel down and stood up. Everything went black and I stood there, trying to remember how to breathe. It's like driving, breathing. The more you think about how to do it, the harder it gets. I stepped blindly over the fallen walls and looked down at my pink tent and thought about the books inside it. I could hear the running river and the wind over the tall grass outside the hall, where centuries of people throwing things out made the soil rich and the wild plants strong.

I am in the infield again, sitting among flowers and tall grasses that grow to my shoulders. I am only a few feet from the little boy, and I can see him watching his uncle and his brother-in-law playing chess. It is early June, one of those weeks when the sun does not set but glides towards the western horizon at intervals that seem much more than a day. It looks like evening because the shadows have lengthened out across the valley until the byre's shade reaches nearly to the water at the bottom of the hill, but no one is going to bed. The bustle of daytime is over. The woman has taken the baby down to the river, saying to the boy that she hopes he might have some memory of the changing light and chiming water when everything is still and dark in the winter. The summer seems endless, but when winter comes, the women and children especially will spend enough time in bed in the darkness to last anyone through the weeks of light.

The men are playing chess outside – no sense in being indoors when the sun is warm on the turf and the ground hums with the business of insects – and the boy sits on the short grass, facing the bay and pretending to play with the boat his uncle carved for him last winter. He isn't playing. He is fascinated by the chess board his uncle has made and particularly by the little people which came from the old country. They are made of a smooth, heavy stone, better than the soapstone the Greenlanders use for such things. You can polish soapstone and it's good to touch, but these shine with a clearer light and they are cold to the skin, even in the sun or by the fire in winter. They look like opaque ice. The farmer is putting his brother's pawns on the ground beside him as he captures them, and the boy edges nearer. When the woman comes up from the river and stops to show the baby to his uncle, the boy picks one up and puts it in his sleeve. He knows enough not to move away at once, as if he has done something with which he does not like to be associated. After a while he crosses the shaded field and goes to sleep in the hayloft, curling up like a mouse in the straw. It is warm and light, but straw is flammable and I watch over him as if I might be able to save him.

I have not had an unbroken night since leaving home. The bad dreams are so vivid I am afraid to go back to sleep, and the cold, dead light of the early hours is not reassuring. I am beginning to feel slow and sleepy in the day, especially during the bright hours after lunch when the morning's work and talk have dimmed my memory of the night. At home, when my thoughts drift through an underwater haze in the afternoon hours, distant from the momentum of starting the day and the pressure of finishing it, I walk the length of the reading room, under the gaze of the dead scholars on the walls and the more or less living ones draped across the desks, and drink six or seven tiny waxed paper cups of very cold water from the cooler outside the ladies'. (Sometimes I meet someone there and we agree that since we're not achieving much anyway, we might as well go to the café. Sometimes I go food shopping and then home to cook with the Afternoon Play, but I expect you already knew that my cooking is not fuelled entirely by devotion to your pleasure. Once or twice I have even bought chocolates from the Maison des Chocolatiers and spent the rest of the afternoon eating them in the cinema, watching Hollywood romantic comedies to which you and everyone else I know would prefer two hours in front of a screensaver. I have found my reputation for diligent scholarship very liberating.)

But all the water is cool here, and the nearest we come to escapism, having escaped, is staring at the sea. I am trying very hard to save the Dorothy Sayers and the Marjorie Allingham until I really need them. So when my tiredness becomes disconcerting, I go down to the river, splash my face and sit on the bank with my feet in the water. They turn blue, but it seems to help. I needed it yesterday. The sun was warm by afternoon, and Ben and I were working in one of the antechambers. We were down to floor level and had found nothing whatsoever, not even a potsherd. There was no wind and the sun warmed my back. I was thinking of you, knowing that whatever I hope you will not water the plants and will then drown them the same day you check my flight number.

And not thinking about what else might be happening back home.

'Nina, you OK?' said Ben. I blinked. His hair shone red in the strong light and the freckles on his face were joining up. There were fewer flowers in the grass than there had been when we arrived.

'I said, are you OK?'

'Yes,' I said. 'I think so. Why?'

'You've not moved in ages. It looked weird.'

'Oh. Sorry. No, I'm fine. Just thinking.'

'Yeah.' He started to dig again, working fast because we'd already decided that the room had been deliberately cleared before the inhabitants left. 'Well, if I was thinking like that, I'd take a break. Ruth has. She went by a while back.'

I straightened my back, which clicked, and rolled my shoulders. 'OK. I will.'

I saw from the hillside that Ruth was down at the river as well, sitting curled on a rock midstream as if she fancied herself as the Little Mermaid. I wondered about heading back to my tent instead, but I knew I'd fall asleep and a nap didn't seem worth the potential trauma. As I came down the path we were wearing between the house and the stepping stones, I saw that she had taken off her trousers. She was shaving her legs with one of those bright pink disposable razors, and there was quite a lot of blood.

'What are you doing?' I asked. Meaning, I suppose, why are you doing it?

'I'm shaving my legs,' she said, without looking up. She was doing the easy bit above the knee at the front.

'Is the cold a good anaesthetic or does it just mean you can't tell when you're cutting yourself?'

'Both, I guess. But I get dry skin, doing it without water.'

There was a bottle of Clinique body lotion on the rock beside her trousers, which were perfectly folded.

'Do you really need to do it at all?' I asked. Mine, I'll have you know, are as hairy now as in the middle of the woolly tights season.

'Yes,' she said, engrossed. 'I didn't get a wax before I left. Dumb, I guess.'

'No,' I said. 'Not dumb at all. I never understand why women spend their hard-earned cash paying strangers to pour hot wax on their bare bodies. I think it's obscene. I mean, does it matter if you've got hairy legs inside your jeans on the west coast of Greenland? Yianni's growing a beard and I don't believe the other two are shaving more than twice a week. Who cares?'

She looked up. It was like making eye contact with one of those brooding, hate-filled sheep. She might, of course, care if she had hopes of any of the guys. I hadn't thought of that.

'I care, of course. It's a matter of basic self-respect, isn't it?'

I turned over a pebble with my foot.

'Not for me. I'm happy to say my identity is entirely distinguished from the length of the hairs on my legs. Grooming is for dogs. If you like your dogs groomed. People could be reading or at least cooking. I don't bother in winter anyway, and in summer I do it on a need-to-know basis if there's a particular reason to wear a skirt.'

'Well.' She went back to scraping away at herself, as if she were hoping to see under her own skin. 'I care how I look. Even here. You can't just let yourself go.'

'Good Lord.' I half-hoped she was Christian enough to be offended. Jim is. 'If I let myself go I hope I'd come up with something more exciting than stubble on my legs. It's not exactly running away to sea, is it?' I stopped myself. What an impoverished imagination.

I turned to go. 'By the way, you might like to check on that Achilles tendon.' Blood was flowing fast from a deep gash above her left heel.

'I have run away,' she said. 'And now I'm here, I'm taking care of my skin.'

'Go for it,' I said. 'Good luck.'

So much for American politeness. It was a shame we were going to share every meal and hear every sound for the next two weeks, but at least I had a sure-fire way of waking myself up. Irritation is almost as good an anti-soporific as retrospective embarrassment. As I left her to her self-lacerations and went back up to the ante-room, it occurred to me that hers was surely not the first blood to flow in that river.

'Better?' asked Ben. 'Does it seem odd to you, working outside?'

I picked up the trowel again. 'Doesn't really feel like work. Because of being outside. I mean, work is something I do on my own with books and I measure it in words on a screen. I like the idea of fieldwork. Having something real to know about.'

His pale hands worked the earth. 'I never got on with English. I don't really like novels. Can't see the point.'

Better, I suppose, than people who confide that they read a book only last summer, what was it, with a picture of a house on the cover. I could see no point in offering a defence of fiction.

'So did you go to America for genuine academic reasons or because you fancied it?'

He looked up. Funny how enviable colouring is so often wasted.

'Well, both. My sister's been in the US for years. Liz. She's married to a scientist in Washington DC. I always liked visiting. The way people don't tell you you can't do things. You know? At school, there was a real sense of who do you think you are, a kid from round here saying you want to be an archaeologist? There's none of that. I'm enjoying it. I like being the foreign one.'

'Surely not the only foreign one, even in the Midwest?'

He put his trowel down. 'I bet Madison's more diverse than Oxford. It's more diverse than Sheffield. Politically as well as culturally.'

'Yeah, right. So they're selling the *Socialist Worker* on campus?'

'They're probably writing the bloody *Socialist Worker* on campus. Seriously. Spend some time in the US.'

I knelt up again. 'I have. And I think America is a good idea that doesn't work.'

He looked at me for a minute. 'Look, Nina, I dare say it goes down well in Oxford. saying things like that, but you're not there now. They're not always going to be too polite to pick you up on it, you know?'

I shrugged. 'Well, you're making assumptions about Oxford, aren't you? I don't even live there any more.'

He turned another scoop of soil. 'Suit yourself.'

There is a grey light over the mountains to the east but the sun has not risen and I am cold, huddled on the rocks above the little house. A lamb calls, the sheep answers. A small rearrangement among the flock nestled under the stone wall below the farm and silence falls again. There is no wind, and I wait.

I see them coming up the valley, moving fast and quietly. Only four men, tall and bundled in shapeless clothes. But they have belts, and knives gleam at their waists in the black and white light before dawn. I want to run and call out. The farmer has a crossbow and the mother and baby might at least hide among the boulders scattering the hill-side above me. The men reach the stone wall and the sheep scatter, but they are used to people and only the lambs bleat a little. One of the men grabs a lamb and his knife flashes. I cannot see clearly but a sheep calls loudly and runs towards the small pale shape on the ground. I know what is going to happen.

Something held my arms and I woke struggling and shouting for help. Fear hammered in my chest and cold surged in my head as I fought for air. I sat up. It was the sleeping bag, of course, just my four-season down sleeping bag, close as a shroud around my upper body. Which did not explain the distinct memory of a cold grasp on my arm, or the rustling in the grass outside too low and slow for wind. The night was dusky, and I sat there a long time, mummified in my bag, listening to something that did not stop moaning and muttering until the sun came back.

It was colder the next morning. Even inside the tent, I could see my breath, and when I came out the canvas and the grass were washed with dew. Catriona was sitting on the wicker hamper, looking through a thin mist towards the silent sea with her artists' block on her lap and her watercolours at her side.

'Can you paint mist?' I asked.

'Don't know yet. I tried, in Skye. Only it didn't come out very well.'

The mist thickened and eddied between us and the beach.

'Did you hear anything in the night?'

She looked round, her hair beaded with tiny drops of mist.

'No. But I was pretty tired. Why, what was there?'

'Oh, I don't know.' I ran my finger down one of the wet guy lines. 'Probably nothing. I just keep thinking I hear things. Only the sheep, I expect.'

She picked up the brush again and dabbled it.

'Sheep can be pretty weird. Ever heard one eat grass?'

'Yes. First night here. I was convinced it was a long-legged beastie coming to get me. And they look at you in a funny way. I don't like them.'

She laughed. 'Still, it's nice to be somewhere where the worst thing you have to worry about is a hostile sheep. One of my housemate's friends got mugged last month, and they slashed her face because she didn't have a mobile phone or a card. I think they thought if they threatened her she might produce some more.'

'Was she OK?'

'Stitches. And the shock. But yes, the knife was sharp enough that they think there won't be much of a scar. The police said she shouldn't have been walking home alone at two in the morning. At least here you could wander about all night if you wanted to.'

I shivered, although the mist was clearing and a watery sun looked over the sea.

'I wouldn't want to,' I said. 'Not for any money.'

I took a cup of that horrible coffee to keep warm, but by the end of breakfast the sun was bright again. I took off my fleece – well, your fleece, it still smells of you, which is both a comfort and a stabbing pain – and watched as Ruth carefully unbuttoned her angora cardigan. It's pale grey and the buttons are real shell.

'Yianni, I'll check my e-mail today,' Ben said. He was looking at the swirl of powdered milk substitute he was stirring into the coffee granules and water in his mug.

Yianni looked up as if he'd heard a hunting horn on the hill behind Ben, and then round at our faces. 'OK,' he said. 'Sure. After breakfast, if you like. But remember to disconnect and compose in Word, then cut-and-paste. We really need to keep that connection to an absolute minimum.'

'Can I too?' asked Catriona.

'Sure. Everyone can, as far as I'm concerned. I thought you wanted to spread it out to get the news more often. Only please don't settle in for a long session with the weekend supplements.'

'It's not the weekend supplements I'm interested in,' said Jim. 'If the first five headlines don't mention the epidemic, I'll stop right there. There was nothing else until the news section, last time.'

I looked up. He hadn't mentioned it until now.

'Is it OK to download and then read stuff off the hard drive, then?' asked Catriona.

Yianni frowned. 'I'd rather you didn't. We're sunk if we get a virus here. I'm not that good with computers.'

'I work on the IT helpdesk back home,' said Jim. 'You've got good virus protection, haven't you? You shouldn't get anything from downloading ordinary files.'

'OK,' said Yianni. 'Just please be careful. We've got no other means of keeping our data here. The computer's wrapped in the towel by my sleeping bag. If you open Explorer it's easy to connect.'

Ben and Catriona went off and I started stacking bowls. Ruth had left half her breakfast.

'Are you finished?' I asked.

She glanced up as far as my neckline.

'You've spilt something there. Yeah, I'm done.'

She got up and went into her tent. Not even someone as self-conscious as Ruth can crawl into a tent gracefully.

I picked up Yianni's bowl from the rock beside him.

'Are you OK?' I asked. 'You look as if you'd really rather we didn't use the net at all.'

He smiled reluctantly. 'You're probably right. Partly, it doesn't seem worth any risk to our work here just to access information we can't do anything about. They only want to know what's in the news out of habit. And partly, I suppose, I like the idea of isolation. It seems silly to come to West Greenland and then check your e-mail.'

'What, you want the full nineteenth-century heroic experience? Messages in bottles and all?'

He shrugged. He was rubbing the rock with his boot and the laces were coming undone.

'Anyway,' I said. 'We can't do anything about the news at home either. At best it's habit, at worst voyeurism. That's why I don't watch it on TV. Nobody needs to watch other people suffering.'

Jim had been reading the Bible he carries around. He reads it after breakfast, and I can't help feeling that it's somehow indecent. Sometimes he closes his eyes and moves his lips and I'm embarrassed to see him. If people want to commune with invisible beings they should do it in private. He closed the book, the pages riffling like money in the wind, and opened his fleece to put it back in the pocket of his polo shirt.

'Are you keeping it there to stop bullets?' I asked.

'Why not?' he said. 'And you need news for democracy.'

'Not much use when the news belongs to big business,' I pointed out. 'And the UK doesn't feel like much of a democracy, anymore. What do you think would have to happen for the people of Britain to make enough fuss to change anything?'

'Poll tax,' said Yianni. 'People only really care about money. If you did it in stages, you could probably introduce public executions of religious minorities without effective popular unrest.'

'What about America?' I asked Jim. 'Is there anything the government could possibly do to make enough people cross enough to take effective action? Because it seems to me you've always been able to do anything at all as long as you keep talking about freedom and the American Way.'

Jim stood up, smiling as if I were a small child showing him how I could stand on one leg.

'Yeah,' he said. 'They vote. And it works. Sometimes it takes a while but it works. I guess you only hear about what we're doing overseas. Foreign affairs don't control our political agenda.'

'That,' I said, 'is abundantly clear. Last time I was in the US, foreign news meant something that happened in the next state. I'll do the washing up.'

I started picking up the plates, which were slithering on the grass.

'Nina?' he said.

'What?'

'Just that—oh, never mind.Leave it. Still having those dreams, huh?'

'Sorry,' I said. 'Sorry if they bother you.'

He handed me his bowl. They didn't really need washing, it was only crumbs.

'Not as much as they bother you.'

I was bringing the plates back up from the river when Catriona came down to meet me. I could see from the hunch of her shoulders that something was wrong.

'What is it? What's happened?'

'The virus. They think it's mutated and it seems to be spreading.'

The sun was still glinting off the sea but the air seemed darker.

'Where? Not in Britain?'

'No. Six cases in Washington DC and one in Carolina. Thing is, they don't know how it got to Carolina. They thought it was isolated in DC. The incubation period seemed really short. If it's longer . . .'

If it's longer we might all have it. We walked back up the hill, the plates rattling in the basket. The shadows of small white clouds scudded over the turf and a seagull ululated from the shore.

'Any dead?'

'Two. One was a nurse who was working with the first one.'

You full of tubes and fighting for breath. You dying. You not there anymore.

'Have they stopped transatlantic flights?'

She shook her head. 'Same rhetoric as last November. You can tell they've only practised for terrorism. Standing shoulder to shoulder and not letting the force of fear turn the tide of democracy. Or something. Every Briton's duty not to panic. Shoulder to shoulder's the last place you want to be if someone's got a viral respiratory problem. I expect your David's right. By next week they'll have decided it's a terrorist outrage. I wouldn't worry, honestly. Not yet. More people die in car crashes in one day than have got sick in the last week.'

'Well, I worry about car crashes too. Is Ben OK about it? With his sister there?'

Catriona shrugged. 'It's a bit hard to tell with all that northern masculinity washing around. Probably? He's had an e-mail from his sister, he says she's not panicking. He says there are a lot of people in DC and probably more of them think they're God than have contracted this thing.'

'Yeah. Only this has more consequences and spreads faster than people thinking they're God.'

We came to the tents and put the basket down.

'I don't know,' said Catriona. 'I bet more people have died of Jesus thinking he was God than died of the plague.'

I laughed. Jim came out of his tent, looked at us, and set off for the barn. I grinned at Catriona but she was red-faced and biting her lip.

'What's wrong?'

'Oh God, I'd never have said that if I'd known he could hear. I feel awful. He must be so offended.'

'Oh, come on. Apart from anything else, it's true. He was just lecturing us on the need for news, which he seemed to confuse with truth, in a democracy. If he likes democracy he has to cope with other people's views. And you're right, religion is deadlier than pestilence. You didn't say Christianity was wrong.'

'I implied it. Shall I go after him and apologise? I don't know what to do.'

'I wouldn't,' I assured her. 'But do if it makes you feel better.'

'I think I will,' she said. 'Oh dear. And I'll be working with him today.'

She scuttled off like Alice's White Rabbit.

Ruth came out of her tent, fully made up. I was beginning to feel as if I was on a stage set.

'Did you hear what Cat said about the epidemic?' I asked.

She looked at my top, which still had the coffee stain on it.

'Yes.'

'Do you know anyone in DC?'

'Not any more.'

What, she did but they died?

'So you're not worried?'

'No.' Her gaze flicked across my face. 'I don't worry much.'

With which sibylline utterance, she went up the hill.

Yianni asked Ruth and Catriona to keep going in the barn and asked me to help Jim work on the midden outside the farmhouse.

'OK. Now you really might find something interesting here, Nina. So go carefully, and call me if there's anything at all odd. Any changes in the soil, in fact, anything that's not obviously a potsherd or a bone fragment. I'll be down at the church. Ask Jim if you're not sure.'

'What do you mean, bone fragment? And what would a change in the soil mean?'

'Organic remains,' he said. 'Come on, Nina, we're talking animal bones here. It's a midden, people don't put dead bodies on middens.'

'What about their enemies? Yianni, I told you before we came that I can't cope with dead bodies. I mean it. They give me nightmares.'

I could see him thinking that a dig is a stupid place for someone who fears the dead, but I did warn him and he did say I wouldn't have to deal with any human remains.

'Not enemies, either. They were living in the house, remember.'

He's right, of course. The Norse have far too strong a sense of the agency of the dead to go round burying people in the back garden. I squatted down and started to dig.

I tried to ask Jim about his thesis. No one here is meant to be writing up, so it's usually a safe conversational gambit.

(How is Daniel doing, I wonder? Last time I saw him, he was, he said, really and finally on track to submit at the end of July, only two footnotes needed checking and a page number glitch sorting out. I told him to leave them and get it in – my MA thesis had two page 47s and nobody noticed, even the internal examiner has better things to do than read the page numbers – but I'd bet a freshly picked salad he's still working on it. Or buggering about on the internet not working on it. He really is proof that you can be too rich, you know.) Anyway, Jim didn't want to talk about it. We worked on for a while, but his sniffing and the movement of his hands distracted me. I didn't know him well enough to share a silence.

'Have you always wanted to come to the Arctic?'

He paused a moment. 'Yeah. Since I was a little kid.'

So at least there was going to be conversation, even if about his childhood. I was expecting right-wing folksiness, all simple hometown values and the importance of self-sufficiency, with a twist of self-deprecating humour if I was lucky. Something blue and smooth began to appear through the soil and I had a moment's excitement before recognising a mussel shell.

'Was it Hans Andersen got you going? For me, it was the illustrations in *The Snow Queen*.'

'Close. My mom was always worried we'd think more about Santa Claus than Jesus at Christmas, so we had more books about camels crossing the desert than Santa in the snow, but there was one I got for my birthday where the reindeer were flying in front of the northern lights. I was completely fascinated. Well, I still am. I saw them in Tromso last winter and it was amazing.'

So you see, it would be worth going to that conference in Finland.

'I'd love to see them. Were they all you hoped?'

'Yeah. They were. It's odd, when they say it's dark all day you expect it to look like night, but it doesn't, quite. It's like the nights here – even when it's light at midnight it doesn't look like daytime.'

He prised a potsherd, a big, curved one, out of the ground and dusted it off. The shadow of a bird flickered over us.

'I thought that was just tiredness,' I said. 'Tell me about the northern lights.'

'I can't. I think it's one of those things you can't describe. You know how writers always compare it to searchlights and curtains of light, but that only makes sense once you've seen it. It's just bigger than either of those, somehow. I can't do any better, not to a literary girl like you.'

He met my eyes a moment. Come, come, Mr Darcy, I thought. You, my love, having nothing to fear from a large Republican from the Midwest.

'I keep half-hoping we'll see them here,' I said, 'but we won't, will we? It's too early.'

'It is. I think sometimes they might be visible in October but there's not a hope in August. Anyway, it's barely dark at night.'

'I know. But it is different. You can tell autumn's coming.'

He glanced out to sea the way we all do at the mention of anything changing.

'Well, it is. But we'll be gone before it gets here. The Arctic's always changing, that's why I like it so much. Even in winter it's not quite static. You can kind of feel the planet moving all the time. It's like having an extra-terrestrial view.'

'You mean a divine view,' I said. 'So you're happy here? You're enjoying it?'

He looked at me as if I'd questioned the hardness of the stones or the wetness of the sea.

'Sure I'm happy. I like it here. But I'm usually happy, I'm happy at home too. I don't think I have anything to be unhappy about.'

Can you imagine that? Honestly? I'd like to attribute this fluorescently good mental health to stupidity, but all the evidence is against it. He's got a full scholarship from Harvard. He might be faking it to succeed, since I shouldn't think you get American grants and scholarships by being neurotic and miserable, which is practically a necessary criterion of success in Oxford, but if so, I think he's kidding himself as well. I was tempted to offer him a list of things to be unhappy about, starting, perhaps, with war in the Middle East and featuring climate change, pandemics, human rights violations, the fallibility of love and the certainty of death, before moving on to the lack of slow-proved bread and good olive oil in rural Greenland.

Down at the church, Yianni was setting up another grid and I put my trowel down and rolled my shoulders while I watched.

'Tired?' asked Jim.

'My shoulders get stiff.'

I let him hold the moral high ground while the frozen bits of my back thawed a little. It's the usual place above the shoulder blades and I wish you were here, among many other more urgent reasons, to knead those knots. I thought without a keyboard it would get better, but a trowel seems just as bad. So much for manual labour.

I went back to work. After a while it seemed that my end of the midden had nothing but mussel shells. Even that restaurant in Southwold puts them back on the beach so I can't see why the Greenlanders kept a heap outside the house, and unless they were very efficient eaters of mussels it must have smelt bad. Not to mention the flies. They had flies, did you know that? Brought like the plague from Norway and, like the plague, died out when there weren't any people left to live on. The mussels look exactly the same as modern mussels and I couldn't see that it mattered very much if they got broken.

'Careful!' said Jim. 'They're still finds, you know.'

'Sorry. I'm just bored. They're all the same and I don't think they've changed in five centuries. Tell me, what made you decide to do archaeology?'

His hands kept moving through the earth at his feet.

'Oh, that's easy. Because you might find something new. I mean, something you hadn't thought of. A real surprise from the past. I liked History but it's just stories. No one really knows what happened and what they say is based on what seems probable now. Archaeology just seems more honest. It's there or it's not.'

'But you interpret what is there. You can read material culture. Well, you have to read material culture. These mussel shells don't mean anything on their own, they're just mussel shells. We read the land and say they're by the house, which means somebody put them there, and we eat mussels so we're assuming the Greenlanders ate mussels, rather than say sacrificing them or bringing a pile of shells up here for some other reason. And they're at ground level so we assume the Greenlanders put them there and it's not that someone came along later and dug a pit and filled it with shells. Archaeology is reading, just earth rather than text. And you could argue that there's less slippage reading words than land.'

'I know.' He shifted but went on digging, pulling out bits of white bone that even I could see came from something smaller than a person. 'All that theory stuff hit archaeology while I was an undergrad. But it does have a scientific grounding, you know. There is a legitimate claim to objectivity. History only tells you what the people who wrote it want you to know.'

'OK. So what, objectively, happened to the Greenlanders? Why do you think they left? You think it was plague, you said, even though the burial patterns are all wrong and none of the lab work has found the virus.'

He held two bits of bone together like parts of a broken cup, and then shook his head and added them to his pile.

'I don't think there was a mass epidemic. I don't think there's going to be a real epidemic now, either, come to that. I think it would be very odd if the plague killed half of Scandinavia and more than half of Iceland but no one in Greenland was even infected. There must have been ships, we know they had external contact right through the plague era, so some farms must have had it. It's perfectly possible that the population was too sparse for an epidemic. No, I think it was climate change. I think the combination of the mini Ice Age and over-farming poor land meant they just couldn't feed themselves and their stock.'

'So they all starved?' The mussel shells seemed endless and I kept scratching my fingers on the sharp edges.

'That I don't know. You'd expect a different burial pattern. Babies and children tend to go first in a famine.'

'So they went to Vinland?'

'Doesn't seem like it. Unless most of them sank on the way.'

'Back to Iceland?'

'No ships. No wood or iron for building ships. And you'd expect documentary evidence. They wrote down everything else in Iceland.'

'Maybe that codex got lost.'

'And none of the others mention it? Maybe. But if you want to know what I really think, I think we still don't know. And I'd like to find out.'

The wind was picking up and my hair kept getting into my eyes and mouth. I pushed at it with my wrist, not wanting soil and organic remains on my face. I was still working through mussel shells, and I remembered that new restaurant, the one I went to with Eva. Fifty different ways of serving mussels, each one a better argument for sticking to the white wine, garlic and parsley approach than the last. I don't think they have mussels in Hawaii, but if they do I am quite sure they don't cook them with pineapple. Not twice, anyway.

'What do you think the Greenlanders ate with their mussels? They'd have steamed them, I suppose.'

He looked up. 'I don't know, Nina, I don't think anyone's done much work on medieval Greenlandic cuisine.'

'Well, I'd have thought it mattered,' I said. 'I don't see how you can postulate that people starved unless you know what they were expecting to eat. Mussels with angelica? I'm going to try cooking with some angelica tomorrow. It won't make those noodles any worse. Some seaweed is edible but I don't know how to recognise it.'

'Stick with what we brought, huh, Nina? Believe me, no one wants food poisoning out here.'

I wondered if there was a Greenlandic version of *Food for Free*, but I suppose Food You Pay For is probably more of a rarity round here. Perhaps if we did up one of these barns we could run a company offering foraging holidays. What do you think? And a little cookery school. We could get Dan to run it when he abandons the thesis. Yianni could come and give lectures on local archaeology. He says he used to do that in Crete, when someone started taking middle-aged, middle-class package holiday people to his grandparents' village.

Jim had made four separate piles of bone fragments, some of them so small I knew I wouldn't have bothered to pick them up even if I'd recognised them. I wondered, with sudden alarm, how many of the white specks you see in any soil are dead things' bones.

Jim grunted in surprise and put his trowel down.

'What?' I said. 'What is it?'

He began to brush at the soil with his fingers, and I remembered Yianni's undertaking that there would be no bodies in the midden. I stood back, ready to look away, and could still see something pale coming through the ground.

'What? What have you got?'

He went on dusting the soil, and then reached for a stubby brush he keeps with his trowel.

'Some kind of polished stone,' he said. 'Get Yianni, will you?'

Polished stone sounded safe enough. I set off across the turf, calves stiff from squatting so long in the cold. I found Yianni photographing the ruined chapel with a camera about the size of a small chocolate bar, which seemed a less arduous way of passing the time than anything the rest of us were being encouraged to do.

'Nice camera. Is that from your grant?'

He went on positioning it, in the peculiar tai chi practised by people stalking a pleasing object with a digital camera.

'Necessary equipment. It'll belong to the department afterwards.'

'Mm. And I bet it will live at your house.'

He put it down. 'Did you come here to demonstrate arts and humanities' envy of social science research funding or was there something more immediate?'

'Jim's found something he wants you to see. Polished stone, he says.'

Yianni put the camera in the pocket of his worn grey cagoule and started up the hill at a fast stride like that of someone pretending not to run away. He'd finished the string and stake grid across the inside of the chapel, but it didn't feel any less desolate for the addition. The walls were still high and the doorway stood as defenceless as an open mouth. It faces

East, of course, towards the Second Coming no doubt, but also away from the sea and more immediate arrivals. There's a fine view of blowing grass and black rock and sky, and meanwhile anyone could creep up. Someone did. There are scorch marks, and I guess there'll be a body in the church that never had a burial. The kind the Norse really didn't like to have around. I shuddered, suddenly sure I was being watched, and followed Yianni up the hill. I didn't look back until I could hear his and Jim's voices, clear and ordinary in the cold wind.

They were standing now, studying something cupped in Jim's hand.

'Look,' said Yianni. 'I think it's marble.'

I peered at the little object, held like a fragile bird on Jim's slab-like palm, and couldn't breathe.

'Could it be Inuit?' said Jim. 'The Norse didn't go for figurines, did they? And where would they have got marble?'

I tried to inhale but my chest seemed tight shut.

'Oh, there is marble. But not green. And nowhere near here. And I'm pretty sure there are no Norse marble artefacts. I don't think anyone's found any imported craftwork that didn't have an obvious ecclesiastical connection.'

'Well,' I said. 'Maybe –'

It was no good. The mountains blurred into low-definition pixels. In and out and in and in and –

'Nina, you OK?'

'Yes.' I still couldn't breathe. 'No.'

'She can't breathe.'

Jim looked alarmed. 'She asthmatic?'

'No,' said Yianni. 'She's panicking. Nina, have you got a paper bag?'

I shook my head. In and out and in and in and out and in and in. My hands tingled.

'She used to carry one,' said Yianni. 'She was getting a lot of these a while ago. It looks bad but it's OK. She won't pass out.'

You and your arms around me and your chest against my face. You waiting for me. The mountains came back into focus. The singed smell of your shirt clean from Iron Maids while I'm still in my dressing gown, planning a bath for when you've gone to work. My shoulders dropped and there was room for air in my lungs.

'Sorry,' I said. 'I'm OK now.' I took a few more breaths. 'Don't know what triggered that, sometimes it's just tiredness. Show me what you've found?'

Yianni gazed at me suspiciously. After one glance, Jim turned back to the polished stone.

'Here,' he said. 'Isn't it a beauty? A little figure. Maybe a doll?'

Ritual or ludic object, I thought, meaning it has no apparent purpose or function. In the days before women did archaeology, Yianni told me in the British Museum, Ancient Egyptian mascara wands and eyelash curlers got categorised like that. (Though the first time I saw an eyelash curler in Hayley Robertson's schoolbag I thought it was some kind of tea-strainer.) The pale green stone was still dusted with soil from the long burial.

'No,' I said. 'No, I think it's a chess piece. But I doubt you'll find the set.'

Yianni looked as if I'd said it was a DVD. 'I doubt they played chess. Other pieces would have been found. You're right, though, it might be another kind of game. Jim, you remember that article on Norse games? Was it Ben's guy at Madison?'

'Vaguely.' Jim stared at the ground. 'I'll keep going. See if there are any more.'

Yianni looked at the midden as if it held the keys to the kingdom, or at least tenure somewhere with ski slopes and undergraduates who know about capital letters. He was almost reaching out for Jim's trowel. I saw him make an effort.

'OK,' he said. 'Of course. You found it. It's yours, really. Keep on. Just let me know the minute you see anything, OK? Mind if I clean this guy up for you?'

'Be my guest,' said Jim.

By late afternoon I had a headache. The sun stayed high and the light was strong, pressing sharp shadows into the grass. I remembered the beach below Yianni's parents' house where we could lie and float under the pine trees, and then coming out of the sea to find his mother threatening to take all the food back to the kitchen if Yianni and his dad couldn't barbecue fish together without the speculative resurrection of the lost Oedipus trilogy. The dust under my nails and inside my ring itched and my hair felt stiff.

'You know, I think I might try a very quick swim,' I said. 'The sea might be warmer than the river.'

Jim sat back on his heels and looked at me.

'I mean, the river's glacial, isn't it?' I went on. 'And the sea's got the Gulf Stream to warm it up.'

'You're telling me you're going to swim in the sea? Here?'

I shrugged. 'I've got a headache. I like swimming.'

'You're telling me you brought your swimsuit?'

'I always take my swimming costume. At worst it's emergency underwear.'

'If you're going to swim, I'm going to watch. In fact, I'm going to take pictures. But I should tell you that the Gulf Stream goes down the east side of the Atlantic.'

I remembered swimming on that black sandy beach in Iceland. It was all right, wasn't it, for a couple of minutes? Refreshing. People are always jumping off docks in the Stockholm archipelago.

Catriona, who swims in Skye every summer, wanted to come too, and she'd brought a swimming costume. Yianni was still exercising the camera.

'Hey,' said Jim. 'The girls are going swimming!'

'Women!' I shouted from inside the tent. 'Why don't you come too?'

I came out. Catriona stood shivering in the sun with a jumper round her shoulders.

'Swimming costumes with legs,' I said. I told you they don't make them just for me. 'From Sunshine and Cloud?'

Goose pimples rose on her arms. 'No idea. Mum bought it for me. School swimming. It's lasting well.'

Thank God for Catriona. My shoulders clenched against the cold.

'Towels?' I said.

'I'll bring them,' said Yianni. 'I have to see this. I have to photograph this.'

I reached back into my tent for your fleece. Well, I bet when people jump off docks in the Stockholm archipelago they don't wander about half-naked for hours first.

'Come too,' said Catriona.

'I will,' said Ben. 'If only to tell my mates. We did some wild swimming in the quarries this spring.'

'Crap reason for doing something. Come on. England expects.'

'Scotland,' muttered Catriona. 'Or maybe Yorkshire.'

Catriona and I set off, barefoot and giggling, across the rough grass. There were stones, but even before we got to the beach my feet were slightly anaesthetised by the cold. Most of the pebbles on the beach were smooth and rounded, and we clutched each other's hands as we wobbled towards the waves. The sea was a dramatic and uninviting black, but at least there were no icebergs in sight. I put my foot into an incoming wave and looked at Catriona. She bit her lip and a gust of wind off the sea mocked our joke. Real cold, not like North Wales or the Hebrides.

'Defeated?' asked Yianni.

I put the other foot in and stood there while a larger, icy wave chucked a couple of stones at my ankles. Catriona did the same.

'Take us a while, at this rate,' I said, trying to stop my lips shaking.

'Mmhm,' she agreed.

I took another step. I couldn't feel my feet anymore but my legs hurt. Then there was a shout behind us and Ben came running down the beach, wearing what looked like a little pair of shiny red pants. Definitely too small for public appearances.

He yelled and ran into the sea, arms waving. The waves broke around his waist and he threw himself in, swam a proper crawl for about six strokes and ran out again. Yianni handed him my towel and Catriona and I stood and stared at them as the North Atlantic washed around our calves.

'That's my towel,' I said.

'You're not using it. His need was greater,' said Yianni. He looked as if he'd won the Oedipal fish-grilling contest.

I took a deep breath and another step into the water. It didn't feel much worse. Catriona's hand in mine was cold and hard. We stepped again. I was dreading the cold water on my breasts, but before I got that far a large wave knocked us both off our feet. I swam while three waves lifted and dropped me and then staggered back to the beach, followed by Catriona. Ben handed me my towel.

'Sorry it's wet,' he said. 'I'm impressed. Race you back.'

It's not a good idea to run in a wet swimming costume in mixed company. I walked back, shivering, at a dignified pace, and when we got to the river I forced myself in. I was so cold anyway it didn't make much difference, and I couldn't face being all sticky with salt until it wore off. Catriona stood watching and Yianni shrugged his shoulders at Jim.

'English women,' he said.

'What about them?' I demanded.

'Mad,' he said, grinning at Jim.

'But interesting,' said Jim. I knew he was looking at my swimming costume. It's probably a federal offence for a woman to have pubic hair to conceal in America.

Later that day, sitting round the stove that Jim and Ben had persuaded Yianni to take down to the beach because he insisted that we weren't allowed to light a fire even on the stones, I wondered how the Greenlanders had seen the sea. A highway, a source of food, a thing of beauty. Especially here, where even now it's hard and unrewarding to go inland over the mountains, the sea brought them everything from plague and terrorists to glass and the latest fashions. Waves rolled pebbles on the shore and a bird piped along the beach. Birds, of course, are here too as well as in Washington. Lots of wild birds.

'Do you think the Greenlanders went swimming?' I asked. 'Do you think they played on the beach?'

Jim and Ruth looked at each other doubtfully.

'It's not the kind of question archaeology's very good at answering,' said Ruth. 'I can't remember much at all about swimming in the sagas.'

'Heroes sometimes swim in Anglo-Saxon poetry,' said Catriona. 'Beowulf.'

I know about Beowulf.

'But here. Do you think they sat here on this beach and paddled in this sea?'

'I don't know,' said Ruth. 'But I think they probably did nearly all their socialising in and around the farmhouses. Maybe the children played down here.'

'The problem is that archaeology has to be more interested in establishing customs than instances of spontaneity,' said Jim. 'I mean, a particular person going swimming, even every day of every summer, probably wouldn't leave evidence. It would only be if someone made a carving or left a record of swimming that we'd know, and then we'd probably assume it was because swimming was important to that society. But I have to say there would be a presumption that people in the Arctic probably don't swim for pleasure.'

'We're south of the Arctic Circle,' I reminded him.

'But you wouldn't guess,' he said. 'Not from your swimming.'

I threw a small pebble at him and Yianni said it was time to clear up and go to bed. The sea and the turf seemed to glow in the slanting light of the low sun and I stood there absorbing the golden light and the diamond sharpness of the hills against the sky. I was sure I would sleep well.

The house is silent again and I walk across the infield and up to the doorway. The sky is low and tinged with green. Rain falls steadily. The sheep are gone and I am the only moving thing in the valley. The door hangs open, askew on its hinges. I step in. I can hear breathing but the house is dark and I can see nothing but vague shapes, too big to be people. I stand still while my eyes adjust. There are broken dishes on the floor and a big pot lies on its side in a pool of something that was probably food. I move forwards into the dimness. I can still hear the breathing, quick and shallow, and I call out but there is no reply.

They are in the next room. The woman has been tied up. I would like to cover her. Blood is spattered across the room. The breathing is in here and as I pass the mother's body, my eyes averted from what I can hardly see but cannot bear to imagine, I realize that it must be the baby. Perhaps the baby is all right. Perhaps I can rescue the baby.

The baby is lying on the floor in the anteroom. It is not all right.

It was grey and still. No wind, but the outline of a hand clear on the canvas inches from my face, and the noise of breathing still although I was – was I not? – awake. I froze. The hand slid silently down the tent. Nothing moved away. It was still there, silent, waiting, breathing. I lay quite still, in out, in out, in out. It was still there and I was still there. I was not asleep. I should have called, screamed, but I dared not move. In and out and in and out. I sat up and the hand was back, higher now, the thing kneeling or standing over me, and I found breath and screamed and then again. The hand went, but a guy line pinged as something caught on it, and rustled away.

'What? What's wrong?' Yianni shouted. The zips on his bag and then his tent whistled and he was coming in.

'Jesus Christ, what happened?' Ben's voice.

'There was something. I woke up and there was a hand. There. It stayed a while, Yianni, it went when I screamed. That way. Towards the house.'

I was shaking, tears came, the big easy breaths that come with crying. Yianni put his arm around me and stroked my hair.

'It's OK, Nina. It was a bad dream. It's all over now, there's nothing here. Look, just the tents and the river. You had a nightmare, that's all. Have some water now, go back to sleep.'

He handed me my water bottle and I drank, cold tracing my throat and stomach like barium.

'Is she OK?' called Jim.

'Yeah. Nina had a bad dream, guys. She's OK. Go back to sleep.'

'Are you all right, Nina?' called Catriona. 'Can I help?'

I swallowed some more water. 'I'm OK. Sorry I woke everyone.'

Yianni patted my back. 'Now, can you get back to sleep OK?'

'Yianni, there was something. Really. I was dreaming but I saw the hand after I woke up. I waited and it didn't go away. There was something there, and now it's gone up to the house. I heard it.'

'There's nothing there, Nina. Really. Go back to sleep. How could anything come here? We're miles from anywhere. We'd have heard a boat. And don't you think the sheep would have made a noise if anyone had come through?'

'Not if it was something that's always been here. Something we've disturbed.'

He sighed. 'There's nothing. Let's get some more sleep before morning, huh? Good night.'

He went back to his tent and I heard him get into his bag and lie down. He sniffed and fidgeted in a way that must madden his girlfriend, but long after he was quiet I lay there, listening for the stranger.

It was three days before Yianni let us check our e-mail again. He kept stalling, as if he knew something the rest of us didn't. It didn't bother me much – having read your mail and knowing that you were OK and missing me and not, I deduced, seeing much of Cassandra, and, best of all, not due to go to the US any time soon, I didn't really care very much if more people in Washington were too hot and couldn't breathe. Naturally one would rather people died of something less messy, but as long as it stayed on that side – well, I suppose this side – of the Atlantic I felt no great personal alarm.

Ben went first, being the most worried. I went down to the beach. I'd been collecting shells. There are lots of shells that have a Fibonacci spiral inside but are almost egg-like to hold, as if incubating one of those apparently impossible seventeenth-century staircases. Either the occupant is inedible or they've arrived since the end of the Norse colony, as we haven't found any shells round the house, but they are pleasing if not useful. I was gathering shells broken enough that I could see the spiral but intact enough for it to be complete. I can't have both, of course – intact is enclosed and exposed is broken – but I like the paradox. There's a poem Donne didn't write there. We're not allowed to take anything away from the beach, but equally there's no one to move or disturb anything left on the beach, so I keep them behind a smooth, black rock above the place where the river fans out across the stones and trickles into the sea. I'm planning, whatever Yianni says, to bring a few of them home and keep them in that glass bowl on my desk, possibly along with one or two utterly insignificant potsherds. I think I'm too shallow for archaeology.

I found a couple of shells, white against the dark stones, and added them to my pile. Future Greenlandic archaeologists will speculate about the ritual or ludic significance of broken shells. The sky was grey but shiny like the inside of a mussel shell and the sea was quiet, waves lapping like little tongues on the beach and the water out past the rocks moving smoothly as fur. And then I saw that it was broken by a dark head swimming, parallel to the little headland from which you can see into the next valley. I sat still but was, of course, too late. Wearing the red cagoule, I was visible as a marker buoy. The head moved steadily nearer, but there were no boats, and I was on the point of calling out when I recognised it as a seal. It came closer until I could see shiny black eyes and the long whiskered snout. A stone moved behind me and I turned to see Catriona.

'It's coming to see us,' she murmured. 'Keep still.' Slowly, she squatted down beside me.

'Will it come out?' I whispered. Seals, like horses, are bigger than seems advisable at close range.

She shrugged. 'They don't, on Skye. But they're used to people. Watch.'

The seal dived. 'Oh. It's gone.'

'Not far. Move over.' I moved and she perched next to me. 'It'll come up again.'

We waited, her cagoule rubbing mine, shoulders touching. Her hair has roughened here and no one, except probably Ruth, is washing often enough.

'There.' She pointed. The seal was barely a boat's length from the shore, watching us intently.

'It feels weird,' I whispered. 'Uncanny. As if it's got something to say.'

'They are odd. There are lots of stories about them. Selkies, people who turn into seals. And back. It's just curious.'

It dived.

'Some early travellers thought the Inuit could turn into seals,' I told her. 'They thought the kayaks were part of their bodies.'

'I know. One or two Inuit got washed up on Scottish beaches, did you know? There are kayaks in some local museums.'

'Still alive?'

'Yes. Though I think they mostly died in a few weeks. Flu, or fevers.'

'There it is.'

The seal was further out now, still looking back at us. If birds and dogs carry the virus, what about seals?

'Thinking of fevers, what's the news?' I asked.

'I don't know. I sort of came down here to postpone finding out.'

'Are you worried?'

'No. Not really. Not yet. I still think it's probably exaggerated to sell news and scare us into accepting more human rights violations. God knows what the US government wants to do next, but I've kind of lost all faith anyway.'

'All faith? In everything?'

'I try not to think about it. I think we're probably the last generation. Don't you?'

The seal dived again.

'People have always thought that. Wordsworth. Homer, come to that. I still want kids.'

'Somebody's going to be right, though, aren't they? They didn't have nuclear weapons in Wordsworth's day.'

'No. But they had the Industrial Revolution. And we've had nuclear weapons for seventy years. Threescore and ten.'

'Mm. I still wouldn't have kids.'

We will, won't we? With starfish hands and heavy heads and snuffly silk faces.

'I don't even really want to wait till I've got a job,' I said. 'I know I should. I don't want to end up bored and over-qualified making fairy cakes and running the PTA. But I really want babies.'

She shook her head. 'No. I want to paint much more than I want children. What I'd really like is to live by the sea and grow vegetables and paint.'

The seal reappeared, a dot on the glassy sea, and a breeze skimmed across the water.

'That ought to be possible. Oughtn't it? If you don't want much money?'

'You need somewhere to live. It's really hard to live off painting.'

'I know. But couldn't you have a croft or something? On Skye?'

'It's hard work. It's not enough to live on and it doesn't leave time to do enough painting to live on.'

'Don't you like your thesis?'

She picked up some of my shells and traced the spirals with her finger. 'It's fine. I think I'll finish it. But I thought it might be my thing and it isn't, really. I should probably have taken the risk and gone to art school. What about you, are you climbing the greasy pole to the top of the ivory tower?'

'No. I mean, I say so. But there's David and honestly, most of the time that seems enough to ask. It feels as though expecting more might jeopardise what I've got and – there's no way to say this that's not hubristic – but if the Fates aren't listening –'

I wriggled on the cold rock. It doesn't feel safe talking about you. I'm not sure I even want you to know quite how much you matter to me. I may not show you this notebook.

'Then?'

'It's all I want. It's all I could ever have hoped. Nothing else matters. And we're four years in.'

'God.' She leant back and looked at me. 'I didn't think that happened. Really?'

I looked out to sea, eyes filling. 'Mm. Well, it does. It happened to me. And now I've come here and left him and I keep wishing I hadn't.'

Tears ran again. It feels as if I never quite stop crying here, not enough for it to be important when I start again.

'Oh Nina.' She rummaged in her pocket. 'I've got tissues but they're very used. Probably from before we left home. Never mind, go and e-mail him. You'll be home in two weeks.'

I had more or less stopped crying when I found the others kneeling around the laptop as if it were an oracle. Ben's face was red and even Ruth looked mildly interested in the unflattering blue light coming off the screen.

'What is it?' I asked.

Yianni looked up. 'It's not good,' he said. 'You won't like it.'

I put my hands over my ears and shut my eyes. 'Then don't tell me. I don't want to know.'

'Nina? Listen to me.'

'I don't want to know.'

Someone put a hand on my shoulder. I opened my eyes. Ben. The laptop was still on, flickering in my peripheral vision, and the sky was huge and white. They were all looking at me.

'Nina,' said Yianni. 'Stop it. Come on. We're all living with this.'

'No,' I said. I met Ruth's gaze and looked away. 'OK. Tell me.'

Yianni glanced at Jim again, gauging what to say. 'The epidemic. It's spread. Several thousand people in the Washington area, another cluster in Charleston and a scattering of outbreaks up the East Coast.'

'On the East Coast.' Wind played over the grass towards us. My voice sounded too loud. 'Not in England?'

Ruth rose from her knees and walked away, down towards Catriona and the beach. Catriona still thinking about her croft and her painting.

Ben moved his hand. 'Things to do.' He followed her.

Jim put his hand on my shoulder and Yianni turned to look into my face. 'There've been a few cases. Some people who'd flown over.'

'It's not spreading?'

You dying, your perfect body coming apart. You, gone.

'It does spread, Nina. They're containing it.'

'I can't deal with this.' I said. But there was no alternative.

The dark time has come. The bodies lie frozen in the outhouse. The snow gives off a faint gleam, and the pit above the beach yawns velvet black. The boys are down there, digging with whalebone spades. A woman robed in woven wool sits like a stone on the bench outside the house, her feet in their skin shoes buried in snow. I curl unseen beside her, so cold and tired that the grave itself tempts me with shelter and the promise of rest. The heap of earth grows, but they cannot dig deep in the frozen ground. There are no ashes or dust here, but ice fusing with ice, death sleeping through the winter until blood runs again in spring.

A howl rose from the dark hill, a human voice lifted in pain. My ear to the ground heard running feet before something dragged past the tents, down towards the new grave on the shore.

'Nina? You having another bad dream?'

Jim's voice, Jim's voice deep like a cello.

'No,' I said. 'I'm not dreaming. Did you hear it too? That cry?'

'You were dreaming. It's OK.'

'It woke me. Jim, I didn't make that noise.'

'It's all right, Nina. Go back to sleep.' Yianni. At last, the others were hearing the Greenlanders too.

'Yianni, it wasn't me. That noise. Wasn't it that noise that woke you?'

'Just a bad dream. Go back to sleep now. There's nothing there.'

'Didn't you hear the dragging? You must know that wasn't me. It went away. I'm still here.'

'Get some rest, Nina. We'll talk in the morning.'

I unzipped my tent and looked out as soon as I woke up the next morning. Yianni was refilling the little stove, wearing fingerless gloves and a surprising woolly hat as well as jeans and fleece.

'So you heard it last night, too?'

'Morning, Nina. I heard you. You're having a lot of these dreams?'

'I am but it wasn't me that howled. It woke me up. Couldn't you hear, it was coming from up the hill?'

He shook his head, intent on the pouring paraffin. 'Sorry.'

'What do you mean, sorry? You think I imagined it? You think I could make that noise up the hill and then drag something through the tents and still be in my tent a minute later?'

'Nina, there's no one here. You know that. No one but us.'

I clenched my fists. 'That's why I don't like these things moving about in the night.'

'Yeah.' He screwed the cap on the stove. 'Well, get dressed and come and have some breakfast. I found some more dried peaches.'

'Yippee,' I muttered. I bet the Greenlanders were at least able to make some kind of bannock or pancake. There is some milk powder, from which in theory – admittedly given somewhere warm enough for it to ferment – one could make something like buttermilk, and there is some bicarb of soda for washing, so if there were flour there could be soda bread. But there isn't, and everyone else seems happy enough to spend all their time digging and eating food out of packets. I still have the Papillon chocolate against a real gastronomic or moral emergency.

While I was dressing I heard Catriona churning around in her tent and then crawling out. The stove had begun its daily purring and I smelt the familiar mixture of paraffin and sulphurous dried apricots.

'Yianni?'

'Good morning, Catriona.'

'Have you thought about what we're going to do? If this, this epidemic, really kicks off? If they close airports?'

'We'll finish our work. What else? If anyone's close relatives get dangerously ill, the insurance covers an urgent flight back, though honestly it would probably be too late. Otherwise, it's just news. And we're safer here than we'd be at home.'

Dishes clattered.

'We are. But I do worry about my family.'

'No cases in Scotland yet. Worrying won't help.'

'I know. But could we check more often? Things could change fast.'

'We'll see. I'll think about it.' Yianni sounded like a parent. 'Could you make the coffee?'

Yianni asked me to work on my own that day, digging across the floor level of a lean-to addition to the barn where he thought hens might have been kept in the summer. The others were staking out a little plateau up near the scree, an area Yianni had been surveying and measuring the previous day. I hadn't asked why, didn't want to know, but there are things moving up there in the night, things we shouldn't disturb. I settled down with my back to the wall and started to dig.

The fire crackles in the hearth. The hall is full, men gathered around the hearth, their words slurred now and laughter louder and more frequent. The children are playing around their feet, rolling pebbles at wooden skittles, squabbling when the game goes wrong. The women, faces hollow on the edge of the firelight, keep their hands busy with spindles and knitting. The cup passes around the men, one of the children jogs an elbow, spills mead and is slapped by an unsteady hand. He squalls, indignant, not hurt. The women exchange glances. Nearly bedtime. I draw my feet up in the sleeping place and lean towards the warmth.

A tall blond man, his words slurred and tumbling, boasts of the hunting journey he made last summer, North beyond the ends of the earth to a strange and beautiful place where the sun circled high in the sky and there were great white bears on the ice and huge eagles in the sky. Someone with a scar across his cheek begins on the stories of Vinland that he heard at Brattahlid this past summer, the land of warmth in the West where there is night even at midsummer and fish run through the rivers like grain poured from a barrel. Silence falls as they think about this, about the grapes that Leif Eirikkson claimed to have found there. I pull the skins up around my shoulders, still cold, and in the silence a strange sound rises and throbs close by the cold walls. The livestock in the byre bray as if it is on fire.

The men seize their swords and push each other down the hall and through the door. The children are quiet, wide-eyed, but the smaller ones whimper and the women pick them up and hold onto them, straining to hear against the wind and the crackling of fire. Silence holds, and then the terrible sound begins to ring through the hall and across the whole valley. It is hard to believe human voices could make such a noise, but it can – surely – be nothing else. It is so low it is felt more than heard, reverberating in our ribs and skulls. It rises and falls and the women and children freeze in the warm hall. Nothing earthly can make such a sound. And then shouts and screams rise above it, the voices of their own men in fear and fury. A tall woman opens the door and I slip out, although I know I am too late. The darkness outside is almost absolute and the cold air fills my chest like water. There may be the outline of the hillside against the sky but there is no moon and the stars are dimmed by cloud. Fewer men are shouting now, and there are whimpers and groans coming from above the byre. The other sound stops suddenly, silence loud in the dark, and I sense beings slipping away into the night.

A faint rustle on the turf, perhaps a flicker of something pale flapping up the hill. Whatever came to the dark hillside has gone. For now.

I was very cold. My hand had set around the trowel, my arm too stiff to bend. A fine rain lay on my hair and face, my stippled glasses blurred the world. I looked at the earth I had not turned and began to cry. I was no longer sure I could manage another ten days here, except that I know I cannot leave. I cannot. I must, but I cannot. Breath came tight again and I gave in to the rising panic, which changed nothing but passed some time until I had to breathe and think again. In and in and out and in and in and in as the fallen stones dissolved and spun around me.

'You OK, Nina?' Jim in his shining armour.

'No. I haven't done any work. I'm not sure – I'm not sure I can actually keep going. I mean, not that there's an alternative.'

He looked around. I was sitting huddled on the wet ground, hugging my knees, my trowel lying at my side. Mist eddied past the stones.

'What happened?'

I looked at him, his sure shoulders and easy breathing.

'Nothing. I keep – it's like dreaming. I'm not hearing voices, I don't think I'm having delusions, I know they're not there. Not in the day, anyway.'

'Stand up. You must be freezing. What aren't there?'

'Nothing. There's nothing there, I know that. I'm so cold.'

I started to shake. My hands trembled.

*

'Come on. Come down. It's lunchtime, we'll make you a hot drink.'

I knew they were worried about me because Yianni lit the stove and expended some powdered milk and cocoa powder making me an ersatz kind of hot chocolate. When I didn't think about Charbonnel et Walker, it was warm and sweet, and the pain of the hot enamel mug in my cold fingers was consoling. The crispbread tasted more wooden than usual, a new kind that snaps into small squares along the perforations.

'Do you not want that?' Ben was watching. A square snapped across, a triangle and a pinch of crumbs.

'Have it.'

'Sure?' He tipped the plate to his mouth and I looked away. Low cloud muffled the mountain and closed off the beach. Rain too fine to hear settled steadily on us and the ground and the stoically grazing sheep. I was so tired it felt like thinking underwater, gliding and bumbling through a silenced world.

'Let's get back up there, please,' said Yianni. 'We've a lot to do now, and it looks like the weather's turning.'

'Looks *as if*,' I muttered.

Catriona caught my eye and smiled. 'Pedant.'

'Should we take the canopies up, then?' asked Ruth. Rain falls on the groomed and the unbrushed alike.

'Not yet,' said Yianni. 'It might clear. We'll get down closer and then maybe put it under tarps for tonight, take a long day tomorrow and see if we can get them all out before dark.'

'All what?' I asked. I knew what.

Yianni sighed and looked away to where the big rocks peered through the mist and then receded. 'I don't know yet.

It's a pit. A big pit. Soil changes suggest organic remains.'

'Did it have one of those holes for holy water?'

He told me about those. When the Norse Greenlanders were too far from consecrated ground to make the trip for a burial, they buried the body on the farm and left a hole for the priest to pour holy water and consecrate the grave when he next passed. Particularly common with winter deaths, which most were, when the darkness made travelling almost impossible.

'No. Surprising, but no.'

'So we're camping by a mass grave.'

'I don't know yet, Nina.'

'What if they died of plague? You don't want to start your own little outbreak, do you?'

He stood up. 'I'm operating in accordance with the University's health and safety guidelines. I have run digs before, you know. Is everyone finished with lunch?'

'I'd like another look at my e-mail before we go back up,' said Ben.

'Later, all right? Let's use the daylight. I'll see you up there, OK?'

After a few paces he thinned and vanished into the mist.

Ruth got up and put her plate on top of Yianni's.

'I'll go help. But Nina? They weren't that stupid. Nobody buries victims of a virulent epidemic near their house and upstream of their water. You really don't need to worry about plague.'

She too faded up the hill. Ben set his shoulders.

'I don't care what Yianni says, I'm going to have a quick look at the *Post*. He seems to think if we ignore this it might go away.'

Catriona ran her finger around the edge of her plate. 'I think he thinks it's not actually here in the first place. And that whether we ignore it or not will only change how well we work here.'

'Yeah, well. It's where I live and I want to know what's going on.'

He ducked into Yianni's tent.

'What if,' I said to Catriona, 'what if it really spreads? What if we can't get back? Ten days is a long time with this virus, right?'

She shook her head. 'Honestly. This isn't the Middle Ages. Remember SARS? Remember the anthrax after 9/11? The bird flu scare? It'll be selling newspapers like there's no tomorrow but I'd still put my money on war rather than pestilence to finish us off. It may even be just an excuse for war. Bet you anything the Americans announce that actually it's another terrorist outrage. A sinister plague from the East. Come on. I need to get back up to Yianni's not-a-mass-grave.'

'It is, isn't it?'

A guy line pinged behind me. Catriona and I froze, eyes locked. I turned my head slowly, shoulders stiff, to see a tall shape moving away from the finds tent into the mist.

'See?' I whispered. 'See that? There?'

She followed my gaze. 'Where?'

'It's gone. Maybe not far.' I clutched her hand, warm and dry in my froggy cold fingers. 'But you saw it?'

She looked away. 'I don't know. Maybe something. I can't tell, in this fog.'

'You heard it?'

'Oh yes. I heard it all right. Maybe one of the sheep?'

Ben poked his head out of Yianni's tent. 'Heard what? Don't tell me you're scaring yourselves. Bad enough with your dreams, Nina.'

'What's the news?'

'Oh, it's spreading. As it would. Though the death rate seems much lower than they were saying earlier. About thirty percent so far, mostly old people and babies. Still. I had an e-mail from my sister. They're all OK.'

'And in Britain?' asked Catriona.

The stones twitched the whiteness and loomed in again.

'Same. Spreading but lower mortality. Mostly in London. Are you going to check your e-mail?'

She hesitated. 'Later. I'll get back to work.'

'Nina?'

'Yes,' I said. 'Leave the computer. Yianni's not expecting me up there anyway.'

Yianni's tent smelt of dirty socks and unwashed hair, and it was dark after the Disney brightness of my pink one. The only book was something scuffed and bent with gold lettering on the cover, and discarded clothes lined the sides. I squatted in the doorway, as if warming my hands at the laptop's fire, and went straight to e-mail. The rest of the country can lie dying the streets for all I care as long as you are all right.

The fact of your message warmed me but I do sometimes wish you felt able to be more open when you write. Without the reassurance of desire, your words have no heat, and I have to remember the way you looked at me when I showed you that dress for the garden party and what we did that last night in the flat. And time has passed since then; it's like looking at old supervisor's comments to convince yourself that the next chapter will be good. You have, I think, seen Cassandra. Have you? I feel very far from you. E-mail works well enough to semaphore your survival but love is not a virtual commodity. I miss you. I need you. I want you. And I write back telling you what fun we are having because I still fear that if you knew quite how much I miss and need and want, you might run for the hills, the very life of late nights in the office and ready meals that you are pursuing at the moment. Perhaps you like it without me? I often think I would like it without me. I could eat mass-market chocolate and sleep in without monstrous anxiety snapping at my feet. At weekends, perhaps, I could go to bed without having put my mind through a mangle for a thousand words of academic prose first. I could have sex without trying to hold my stomach in all the way through. Other people, I know, live with themselves in these circumstances, and some of them are less productive than me and have bigger stomachs than I do. I could understand how you might find the company of such a person, in the abstract, rather restful.

Back up the hill, stayed by the knowledge that you had at least been all right earlier in the day, trying not to think too much about you still using the Underground and still seeing people when you could be avoiding their germs, I tried to work. The mist curtained the fallen walls like a tent, muffling the voices higher up and cocooning me against their discoveries. I found an oversized bone bead and congratulated myself on identifying a loom-weight. I brushed the damp dirt with my fingertips and weighed it in my hand, wondering who had touched it last and what she had woven, what kind of bone it was and who had carved it for her. Some animal seven hundred years ago sacrificed to the need or greed of the last people to tread the ground on which I sat. They must have treasured their cloth, here, not wearing skins, dressing like civilised Europeans although within a few days' sail of the polar ice. I slid it into a plastic finds bag and noted the date and grid reference in writing distorted by cold. Even in good moments my fingers feel numb.

Most of the women are out on the hillside now. It is not easy for them to find silenced men in the dark. The wind is loud in my ears, and the little rush-lights are soon extinguished. I pick my way back to the open door, where there is a faltering glow and the sound of children crying. I slide around the edge of the room, hoping for a place by the fire while the children howl and their mothers are out in the cold night.

Working in pairs, the women begin to drag the men in. They are all dead, the ones who jostled and boasted over their mead earlier in the evening; the hand that slapped the child hangs limp, half-severed from the arm. The women work silently, their tunics stained red. They do not cry, and the children also fall silent, all except one who begins to scream on a note that fills the high hall. She stands at her father's side, pulling his bloodless arm. The women work on. The ground is frozen hard and there will be no burial for these men until spring. They wash them with water from the pot on the fire, and then bring linen from the chests against the wall. When the men are wrapped like winter babies, they carry them, one at the shoulder and one at the knees, the bodies soft and heavy as blankets wet from the river, to the outhouse where they must await the returning sun. After they have taken six men, they stop and arrange the others by the door. The children have gone to bed, but the women sit and watch their dead, listening for noises on the wind, until the darkness outside thins and the wind softens.

'Good morning! Nina, wake up. Come on, time to get up.'

Yianni was peering in through the flaps of my tent.

'What?'

'It's morning. Why did you sleep with your tent open? You must have been freezing.'

I sat up. My torch, which I keep by my side, was in the doorway, and I couldn't see my boots.

'I didn't. Don't you remember, I went to bed early? You saw me close it. And my torch has been moved and I can't see my boots.'

Yianni looked at me. His eyes were red and puffy and his stubble was turning into a beard.

'Have you been sleepwalking?'

'No. I don't sleepwalk. And even an experienced sleepwalker would be challenged by this sleeping bag.'

Catriona's tent began to move as if there were an animal inside trying to get out.

'I heard someone moving around,' she said. 'Weren't you going to the loo?'

'No. Frankly, I'd rather wet myself than stumble about out here on my own in the middle of the night. No. I keep telling you, there are things here. Something opened my tent and moved my stuff.'

I shivered. Yianni handed me the torch and I didn't want to

touch it, thinking what dead fingers had gripped it while I slept. And the torch was by the jumper I fold for a pillow, so anything that reached it climbed over me first. I hugged my shoulders, feeling sick.

'It must have been you, Nina. Sure you didn't go to the loo in the night and forget about it?'

I looked at him and then drew my knees up and buried my face. There was no comfort, no comfort at all, in being right.

Catriona put her head out of her tent.

'I did hear something. I can't say it sounded in the least supernatural.' She raised her voice. 'Was anyone up in the night?'

'What?' Ben, sounding as if he had his mouth full.

'Were you up in the night? Someone opened Nina's tent and moved her stuff.'

He yawned loudly. 'If I had been up in the night I wouldn't have been moving Nina's boots. Do you walk in your sleep, Nina?'

'She says not.'

I raised my head. 'Can someone please find my boots, anyway?'

'I'll have a look.'

Yianni began to cast around and I watched wearily. Grey cloud blanketed the sea and the turf was wet again. I put my head back down on my knees and thought of you, getting out of bed, throwing the duvet back and forgetting to shake it down again, humming and bumbling on your unnecessarily gradual way to the shower. That mysterious pause you make between turning the water off and the squawk of the shower door opening, often several minutes long. I tried to think of anything I would not give to be in our bed, annoyed by the duvet-throwing and bumbling, and hoping for a cup of tea and perhaps one of those sudden, urgent encounters fitted in between your shower and your suit that leave me so pleased that there seems little point in getting up.

'Nina? I can't find them. Are you sure they're not in your tent?'

The cloud had not shifted, but a seagull flapped out of it like the albatross and landed fussily on the rock behind Ruth's tent.

'How could I possibly lose a pair of muddy climbing boots in a two-person tent containing one person and nineteen books?'

'Just checking. OK, guys, can everyone be getting up, please? Come on, we've less than two weeks left now and we'll be losing some days to the weather. I want to get that pit cleared as soon as possible. Ruth, Jim?'

'All right, Yianni, we're not asleep. Hadn't you better find Nina's boots? She can't work barefoot.' Ruth sounded as if she'd already been to the gym, showered, chosen her outfit and put her make-up on.

'Speak for yourself,' said Jim. 'Yianni, it's not six yet.'

'We've a lot to do. Breakfast in ten minutes.'

I breakfasted, or at least sat on a rock pushing prunes around a dish, with my feet wrapped in Catriona's scarf. Yianni sat there drumming his spoon against the rim of his bowl, his impatience visible as a shadow. As soon as Ruth popped the last dainty mouthful past her lipstick, he stood up.

'Right. Nina, I want you and Ruth up at the pit today. There's some unskilled digging but I want an expert eye up there. Catriona, you can go on with the levels in the chapel. Jim and Ben, catalogue finds until Ruth needs you up there.'

'I can't,' I said. 'Yianni, I really can't. You promised me, no bodies. You promised, remember, that night at the flat?'

'Once you get to the bodies, Jim and Ben will take over. But I need you up there today. We're running out of time here. There's no guarantee of another season's funding and anyway it'd delay publications by a year if we have to come back. I told the ESRC we'd do it in one and we will do it in one.'

My chest tightened. 'I can't cope with dead bodies. I'm sorry but I can't.'

'You don't have to. But I expect you up there as soon as you've found your boots.'

It doesn't help me act like a grown-up, being spoken to like a child. I started to cry. 'I can't work barefoot, anyway. I'll get frostbite. I need my boots.'

He slammed his plate down. 'I'll find your fucking boots, all right?'

'Hey.' Jim stood up and put his hand on Yianni's arm. 'Cool it. We're working together here, remember? No one's trying to waste any time. Come on, I'll help you find the boots. Girls, you get the dishes, OK?'

They walked off up the hill.

'Girls?' I still couldn't stop crying. 'Girls? Screw the pair of them. Get the dishes. Smug bastard. Bloody fundamentalist.'

'Yeah,' said Ben. 'I'm not a girl.'

Catriona and Ruth began to laugh and I turned round and crawled back into my tent. I burrowed into my bag, pretending to believe that, without boots, I'd be allowed to spend the next week curled up there, reading *Middlemarch* and waiting for time to pass until I could get back to you. I didn't even really feel like reading anymore. I pulled the bag's hood close round my face, rolled into a ball and shut my eyes.

I open the door of the outhouse, which creaks on leather hinges. Outside, the snow shines under the moon, but in here the darkness is as thick as wool. I can see toes poking through frayed linen in front of the door, but otherwise the dead wait in hiding. I move forward, feeling rough cloth covering something firm and cold even to my cold fingers. The dead are standing, frozen stiff and stacked against the walls because there is not enough room for them to lie on the floor. My fingers find the papery smoothness of a frozen face, the prickle of eyelashes and then the open ice-cube eye.

'Nina? We've found your boots.'

'Nina? Have you gone back to sleep?'

'I'm awake.' I looked up. Catriona.

'Come on. You must have been sleepwalking, they were up by the pit. Have you been dreaming about it?'

I sat up. The boots were muddier than I'd left them. 'No.'

She held them out and I took them reluctantly.

'Come on. Yianni's waiting. He seems a bit calmer now he's up there. I'm sure he'll let you stop as soon as Ruth gets close.'

'Close to the bodies.'

'You won't have to see them. No one's trying to traumatise you. Though honestly, they're just bones. We've all got them inside us.'

I shivered. 'Don't. I'll come. I'll be up in a minute. I just keep thinking, what feet have been here?'

'You think Goldilocks has been trying on your boots? Nina, you must have been sleepwalking. No one else could have come into your tent without waking you. And I really can't believe in a ghost that borrows people's boots. Don't ghosts come with footwear?'

'I don't sleepwalk. I never have. I couldn't even do up my boots with my eyes closed.'

She was looking up the hill, still holding her trowel in one hand.

'Anyway, you might as well come and do some digging.'

'I'll be up in a minute.'

It took me a long time to come out of my bag and put the cold boots on. In the first few days here, the muscles in my calves ached from the weight of wearing boots all the time. Now I suppose shoes would feel strange, but I wanted to be back in the world where I choose which shoes to wear each morning. I remembered the purple pair with embroidered flowers I bought in the sale that last week at home, still waiting in their tissue paper. I think they're one of the brands that come with a little cloth shoe-bag, one of those pointless but acute pleasures like the layers of boxes and bags around real jewellery. I am, by the way, still wearing the ring and doing my best not to scratch it. It seems to be getting rather loose.

*

When I got up to the site, only Ruth was there, crouching in one corner and digging as if she were making a burrow.

'Sorry I'm late.'

I sat on the edge. The pit was deep enough that I could dangle my legs, a neat rectangle of rich black loam in the coarse grey turf.

'Do you want to start over there? It's just digging at the moment, just getting deeper. Tell me as soon as you see any changes in the soil.'

I sat still. The sky was low and grey over the hills, but the sea lay dark and clear.

'Start in that corner?'

I stood up and walked around the edge, not wanting to stand on the dead.

'Nina, they won't bite. These bones.'

I remembered the teeth. 'They did. Once they bit.'

'When they had muscles. Can I ask you something?'

I sat down again. The sheep in the field below rose and then settled like birds. 'Why did I come on a dig?'

'That's the one.' She hunched away from me, working fast.

'I like Greenland. I'm interested in the Greenlanders. I suppose, mostly, I don't believe in passing up experiences.'

'And you didn't think that excavating burials might be part of the experience of archaeology?'

'Yianni promised. He promised me.'

I sniffed and swallowed. My nose ran and I wiped it on my hand. I pushed myself off the edge and into the pit. Squatting down, I began to move soil around as if it were food I didn't want to eat. We were sheltered from the wind, in the grave, but I could hear it moving over the grass. When I looked up, I could see the mountain and the sky but not anything low to the ground at the side of the pit. The soil darkened down the side by my face.

'Are you actually doing anything there?' Her face was pale, the freckles dark.

'I'm digging.'

'You're not digging, Nina, you're wasting time. Time and research money. Yianni could have had someone who'd work, you know. Another archaeologist. This is research, not tourism. This isn't Nina's gothic tour of Greenland.'

I started to shake. Breath came fast. I turned away from her eyes, back to the cold earth.

'You've never lost anyone, have you? The dead don't borrow your boots, Nina. They don't come visiting in the night. They're gone, for always, and you don't see them again. So shut up and dig, OK? And stop crying. You have nothing to cry about, with your perfect relationship and your rich boyfriend and your little ring that you never stop twiddling. You know fuck all about death, Nina. Fuck all about dying and losing the people you love.'

A high whine, like a military plane, sounded lower and louder in my head. I curled up on the soil and rocked, waiting for her to stop, and when she didn't I climbed out of the pit and came back down here. I am very sorry to let you down, but I can't keep going anymore. There are dead people in my mind and rotted flesh on my fingers. Open mouths and clawing hands. Bone feet in my shoes and crying voices. It's dark here, cold and dark, and strange sounds come carried on the wind.

RUTH

You were right. I shouldn't have come. I thought Greenland might help. I thought wilderness and unspoilt scenery might cure me where you have failed. I thought I might look at the ice and somehow feel better, be free of my widow's hood, but the healing power of natural beauty turns out to be one of those popular myths. Like counselling. Like closure. Like God. People told me I'd find closure after the funeral, as if it's a motel on the highway through the Stages of Grief, and then you said it can begin when people stop feeling angry. Well, I'm still angry, and I'm angry with you, and I'm angry with them. The pursuit of happiness is a right, not a responsibility. And you made my grieving sound like an addiction, something to be cured with a twelve-step plan, sitting there among your kilims and your abstract splodges with your kids racing their cars up and down the hall. Telling me how to mourn while your wife, your fragrant wife, stirred her soups and practised her fucking piano. Stay on the interstate and you'll come to the border with Acceptance. Americans love spatial metaphors, have you noticed? We can make anything sound like going West in a covered wagon.

RUTH

I'd always been the American One. In Paris, in Amsterdam, I wore my nationality rebelliously like an expensive dress, but when I got into Columbia I had to come home and take it off. I thought it would be a relief to reclaim my heritage. For a while I was right. I didn't have to explain myself, no one was surprised when I could speak the language. I started to find my way round the city again, and then I met Lisa and Kate, and then James. By the time we moved into the Plum Street apartment, after Christmas, my accent was at least in sight of land and I'd stopped telling the truth when check-out guys asked me how I was doing today. Fine thanks, and you? Then – afterwards – I found that there's no place here for death. We have to call it a problem and go to trauma sharks like you to get it solved. Counselling. As if 'counselling' can make death better. As if there's a software patch for mortality. You can have as many hyphens in your identity as you like, no one raises an eyebrow when I say I'm a French-American raised 'in Europe' (even you, with your hyphenated Germanic pretensions, think of 'Europe' as one place), but I'm not allowed to be a grieving American, an American who can't face her own apartment because of the person who's not there, an American who still has her partner's voice on the answerphone when he's been gone for months and calls it from the sofa on her cellphone just to hear him speak. An American in whose mind runs a constant film of someone burning to death.

Sorry. You don't want to know about that, do you? You don't want me to keep thinking about that but I do. Still. Despite what you kept telling me while the sunlight lay purple and red and blue on your floor and Brahms tinkled and you ignored the phone in the hall and passed the Kleenex. People's eyes explode when they burn to death, you remember that? I assume you'd be unconscious by then but what kills you when you burn is blood loss. A person can't die of pain, which means there's no limit, in time or intensity, to the pain you can feel. I can't leave him there, can't turn away from what I cannot share. At least we know how long it took. I find little comfort in the knowledge of those forty-two minutes between the 911 call and the hospital's decision that he was dead on arrival. There must be a limit, mustn't there, to how bad forty-two minutes can be? And I'm still wondering, was he alive, was his heart still beating, when the firefighters dragged him out? Did he die in the ambulance? Did he know? Anything? Was he still there, still here? His dear face. The police have photos, which they won't let me see. At the service I wanted to throw his mother's floral tributes at her head and open the coffin like a piano, show everyone the broken thing James had become inside the shiny box with its wedding-cake topping.

Sorry. You don't want to know about that. You think I should 'tune out that voice'. Tell James what I need to tell him and understand that time passes. Well, that voice is mine, it's the *uber*-voice, the voiceover, of every breath I take. You tried to make me tell another story but I'm talking now and you can't answer back, you with your perfect life and all the answers. With your blond hair and your suntan and your statuesque feet in your Volkschue sandals. I used to sit there, you know, and look at your toenails, like little pink shells, and the way the hair grows straight and shiny on your legs. Most guys have body hair like pubic hair, wool not satin to touch. Most guys cover it up at work.

Anyway. Greenland. I suppose it is beautiful, in a bleak way. The others seem to like it. Catriona keeps mentioning the light and Nina seems to take pleasure in the shore. There were flowers, which I hadn't expected, but they've wilted and died now and it's getting colder day by day. The digging is good, I can still work – so you were wrong about that – and it makes me tired which means that hours pass at night without me having to endure them. But I don't like it here. You wouldn't think that in the midst of missing James I'd have any longing left for plumbing and heating and doors that close, but I do. I expect you think that holds the seeds of my recovery, that one day I'll think so much about flush toilets and ceramic hair-straighteners that I won't think about James, when the truth is I'd shave my head if it would give me a few more seconds in his arms.

Sorry. I came here to think about Greenland, didn't I? There are six of us. Yianni is running the dig. Greek. British-Greek, if the British did hyphens, which they don't. Catriona, Scottish girl who paints and doesn't wash. Nina, British, blonde, neurotic, literary critic not archaeologist, much indulged by Yianni who has a crush on her. Jim, nice Christian guy from Iowa who somehow pitched up at Harvard. Ben, from the north of England via Madison. We are living in a circle of tents in a field next to the site and none of us has been more than a few hundred feet from this circle since we arrived. The others, except Nina, are working on the Norse Greenlanders and really care about this dig. I keep thinking I shouldn't be here, running my film of burning and fantasising about hot showers and somewhere to cry unheard, but after I gave up on you I knew I had to leave New York. I still couldn't face the apartment, James not tipping his chair back in the kitchen and not leaving his shoes for me to fall over on the mat and not having the radio on loud in the shower. Not leaving Sur le Vent from duty free in a ribboned box on my bureau or Roussillon underwear beneath the pillow. The last few weeks, I was staying in the library till it closed and then hanging about in Barnes and Noble until words blurred on pages and I couldn't remember what I was reading. Then I walked back to Plum Street, almost hoping for a trigger-happy mugger, and went to bed in the dark so as not to see that he wasn't there. I dreaded sleep in case I turned to fold myself around his back. Some nights I woke, found he wasn't there and padded through to the lounge to tell him to stop sleeping in front of the TV and come to bed. Mornings were a long time coming. At six I went straight to Starbucks for coffee and a muffin and then back to the library, walking to fill the time until it opened. There are a lot of people spending too much time in Barnes and Noble and I was getting to recognise them.

I'm not really very interested in the Norse Greenlanders. It was Prof Ekstrom's idea to send me here, trying to save my doctorate if not my mind. (And he's right, you know. The doctorate is worth saving. Research has a value independent of its perpetrators, and independent of your view that psychological integrity is the only thing that really matters. It doesn't. You can be happy and boring and stupid and self-absorbed and you wouldn't have to look far for an example.) It's not exactly a dynamic culture – that goes on doing the same thing for five hundred years until everyone's dead. They died of conservatism, going on trying to grow grain because it had worked at the beginning, before the temperatures fell in the 1400s. There were too many of them to survive that degree of climate change but I don't see any evidence that they tried very hard. They'd have needed to learn from the Inuit, of course, and probably ditch their big farmsteads with the volcanic underground heating and streams diverted through the kitchen. Some of them even seem to have had saunas when they should have been saving every scrap of fuel for cooking. What you can grow in Greenland takes a lot of cooking. The Greenlandic Inuit were nomadic because that's what the land can sustain and those Norsemen, arriving at a time when British monks could grow grapes, never grasped that what they thought was normal was the warmest it had been for centuries. They must have seen the winters getting longer and harder and the harvest smaller year after year, and I'm sure some of them bailed out, back to Iceland or west to Vinland, but many of them seem to have hung on. People have found dogs butchered for food and middens full of limpets. Can you imagine trying to feed your family on limpets? Your kids, with their hair even shinier than their shoes. My guess is most of them just starved to death in the end, though only anorexics and lost explorers really die because they don't have enough calories to power a heartbeat. If you can get to that point, it's probably a pretty good way to end things. Most people don't. In famines they die of dehydration from diarrhoea, communicable diseases and maybe vitamin and mineral deficiencies.

Not as good as heart failure but better than burning. Nina has nightmares about the starving ones but I can think of worse. Can't stop thinking of worse.

The archaeology, I will admit, is suggesting more interesting possibilities. We haven't made much of a start on the chapel yet, but there are signs of burning, which would seem to fuel the pirate-raid theory. And the burial pit, apart from having tipped Nina over the cliff-edge of psychosis on which she appears to live, has the potential to complicate recent thinking on Norse–Inuit relations. It's on the hillside above the barn, which is strange given the proximity of presumably consecrated ground around the chapel where they could have had an orthodox burial, and it's big but not deep, unless there are more burials under the ones we've found so far. Yianni's anxious that we might not finish the site this season, and also, I think, anxious either that the epidemic scare might mean people want or need to leave early or that it might be hard to get back next year. He's pretty much leaving the pit to me, anyway, and there's some peace in having a kind of room of my own, even if it is a grave.

Nina was helping but she bailed out the first afternoon, before we'd uncovered the first skeleton. She's been screaming and shouting in the night since the beginning, having nightmares that seem to get worse, or at least more disruptive, as the days go by. You'd love it, a pretty Oxford graduate with fully-paid-up delusional fantasies. Plath on Prozac. It's bad enough being dragged out of dark and silence into the cold light of night by someone else's terrors, but much worse when she expects us to interact with them too. 'There was someone there,' she keeps saying. 'Yianni, I heard footsteps. Yianni, I heard breathing.' Half your luck, I think. I hear no footsteps and no breathing. No feet and no lungs, no rising ribs, no ticking heart under the quilt beside me, no hair feathered on the pillow, no arms to reach for in the night. If sleep is so bad she should stay awake and think about going home to her fiancé who has, she says, already booked the restaurant for her first night back. Remember I told you James had had me make a booking at Pablo's for that night? I didn't tell you they called while the police were still with me in the apartment, to say they weren't opening 'out of respect'. Oh OK, I said, of course. Thank you. Respect for James, I thought, not thinking. And are you OK? asked Juan, as if he already knew. It was the first time I had to tell someone he was dead and I held the phone and stared at the policewoman, the striped tabs on her jacket, the dandruff on her shoulder. No, I said, not really OK. Not great, not today, because the thing is. Well, the thing is, it seems that James. James is dead. Then I handed the phone to the policewoman and stood there, having no reason to move.

Sorry. Self-torture, they say, the few people who kept listening after the first few weeks, meaning self-indulgence. Move on to the next stage, keep up with the American grieving schedule. Nina came and sat on the edge of the burial pit, swinging her legs like someone fishing off a dock while I was working down to the level of the first burial. Come on in, I thought, the water's lovely. I knew she was scared of bones.

'It's only digging at the moment,' I told her. 'Tell me when you see changes in the soil and I'll take over.'

She looked at the sky and didn't say anything.

'Start in that corner?'

She walked around the edge as if it were full of snakes.

'They're only bones. We've all got them. They won't bite.' Or walk, or eat, or kiss, ever again.

'They did,' she said. 'Once they bit.'

He had shaved that morning, taking up the bathroom when I needed to go meet Ekstrom. His jaw smooth, the cinnamon cedar aftershave and toothpaste on his breath, a nick on his top lip that I kissed before he left. Two hours later it was smashed across the dashboard and burnt to fine ash.

Nina walked off and I went on digging, making my way through increasingly friable loam. The wind got colder, whipping my hair into strings I knew I'd be too tired to sort out. It's too cold, now, to wash in the river, and even when I do manage to wash my hair in the bucket it's still damp in the morning. I kept going as the clouds darkened and the wind rose, sheltered by the grave. I would have buried him in his Italian suit. Linen rots slowly and I would have liked to think of it there with him as flesh fell from bone, but his body was no longer the shape of clothes. He bought that suit in Rome, carried it home as hand luggage as if it were myrrh for the baby Jesus. He'd only worn it twice, Kate's wedding and Joel's bar mitzvah. It didn't seem worth the money to me – money Kate or I could live on for weeks – but I guess a good suit never dates. I could give it to Paul but I won't. It lives in the wardrobe next to my red dress, in memory of our dancing at Kate's reception.

I was working across the pit but the soil in one corner was noticeably darker and more crumbly than the rest. I stayed there, smoothing the soil away as if I didn't want to wake him, and by the time I came to something long and smooth rain was beginning to drizzle onto my stiff shoulders. I covered it again, patting it like a sandcastle, and went to find Yianni. Bones, unlike the living, need to be protected from the rain.

Yianni was sitting in the entrance of the finds tent with the laptop balanced on his crossed legs and a notebook at either side. He looked up.

'Found anything?'

'Somebody's long bone. But it's raining. Do you want me to get a canopy up?'

He leaned forward, peering round at the sky. Colour was already fading from the hills. 'Tomorrow. It'll be dark by the time we've got it up.' He looked the other way. 'Is Nina up there on her own?'

'Of course she's not. She left hours ago.'

He put the laptop down and crawled out.

'I haven't seen her. I've been doing this a while. The battery's running low.'

'Shame it's not wind-powered,' I said. You'd think he'd have realised that solar power is not an obvious technology for Greenland.

'I don't think the grant would stretch to a turbine. Did Nina say where she was going?'

I shook my head. He went over to her tent. She pays more attention to that pink tent than to her clothes.

'Nina? You in there?'

'Perhaps she's down on the shore,' I said. 'I'll spread some tarps in the pit, shall I?'

'Please. I'll come up and look in a minute. I just want to check Nina's OK.'

I took my time, smoothing out the crackling plastic and foraging for stones to weight it with. Then I arranged the squares as if piecing a quilt. I haven't done any sewing since James died, not even for Lisa's baby. He must be crawling now. I remember telling Lisa James was dead and then realising that she'd called to tell me she was pregnant. I used to say, tell me how you are. No, really, I do want to know. Better to have something to think about. But not that. Not that Lisa and Jake are making babies James will never see. He liked babies, but he was great with older kids. He'd have liked yours. When we took Bradley to the Science Museum last year, the two of them went off together, pressing buttons and taking turns on the consoles. I snuck off to the café with a copy of *Le Figaro* that I'd found at the kiosk in the park, and when they came to meet me Bradley talked about the specifications of spaceships for nearly forty-five minutes while I ran my eyes over the back page and James made sketches for him. We would have had babies, later, when I'd got tenure and he could cut back at the bank. I'd have swapped the Pill for folic acid and given up the brie. There was no sex, the last week. The night before the accident I stayed up late working on my notes, and when I came to bed he'd given up waiting and was already asleep, his finger still marking *Travelling Light in Kashmir*. But at the weekend I'd missed a pill, nearly took the chance and then remembered Kate's termination and didn't. And that was it, James's DNA erased like the smallpox.

I vaulted onto the side of the burial pit and stood up. Down by the camp, the paraffin light glowed, and the stove flared close by. Figures moved about, their shadows flaring huge and then tilting away across the tents. The sea was dark and quiet and a grey light cracked the horizon. I set off towards the lights, picking my way over the dark turf.

'Tucked up for the night?' Yianni was stirring a steaming pan of what looked like water and smelt like compost.

'Yes. But we'll need a canopy tomorrow. Unless you're sure it won't rain.'

'We'll need a canopy. Ruth, when did Nina leave you?'

I shrugged.

'I didn't check the time. She didn't stay long.'

'Did you see where she went?'

'I was digging. The pit's too deep to see out.'

Catriona put a handful of spoons on top of the pile of plates at Yianni's feet.

'She's not in her tent,' she said. 'And I've looked on the beach. And around the chapel. It's not as if there are many places to go.'

'Was she upset when she left?' asked Yianni.

'She's been upset all along,' I pointed out. 'She's scared of bones.'

'She's a bit fragile.' He stirred the water again and glanced up. 'Was she distressed?'

'I guess. She certainly didn't want to be there.'

Catriona clattered the spoons on the dish. In the low light the fronds of her hair waved like snakes around her face. 'I'm worried about her.'

'Where was everyone else?' I asked. 'Wouldn't you have seen if she'd walked into the sea?'

'No,' said Yianni. 'Supper's ready. And if we haven't seen Nina by the time we've finished I'm going to call the coast-guard on the satphone.'

He ladled out compost water, and then fished for the inevitable noodles.

'Chicken noodle again?' asked Jim.

'There's chocolate pudding.' Yianni passed him a plate.

'Maybe it's a good thing Nina's not here,' muttered Ben. Last time we had 'chocolate pudding' Nina read the ingredients aloud and said that if it were produced in the UK it would have to be described as chocolate-flavoured pudding because it had no cocoa content whatsoever, and added that if Yianni had brought dried eggs she'd have been able to make us pancakes. So could I, I thought, lacy crêpes the way Papa taught me when we were living in Saudi and boredom drove me to elaborate kitchen recreation. The first time I cooked for James I flambéed them, proper *crêpes Suzette* made with a bottle of Cointreau that left me living on pasta until the next scholarship cheque came through.

We started eating. Ben struggles with noodles and I kept my eyes averted.

'She wouldn't walk into the sea,' he said, 'would she?'

'No,' said Yianni. 'I don't think so. At least, not unless something happened.'

'You don't think she's been seeing things again? Does she – you know – hear voices?'

'No,' said Yianni. He put his plate down. 'She's not sick. I wouldn't have invited her. I know she had some depression but she's been fine for years. I mean, she was off medication by the time I met her and that was before she got together with David.'

'Well,' I said. 'She doesn't seem very fine now. What was it with the boots?'

'Sleepwalking, I guess,' said Yianni.

'Unless we believe her,' said Catriona. 'I mean, I did hear something one time. A guy line. She said she saw someone and maybe there was something.'

'Don't you start,' said Ben. 'Don't let her get to you. It's too easy, here on our own. You'll have us all chatting to the Greenlanders. I bet you wouldn't have noticed anything if she hadn't told you about it.'

'I heard the guy line.' Catriona stirred the greenish water in her dish, witch-like in the dark. 'I suppose it could have been a sheep. I just can't think where she'd be. It's not as if she's gone to the library or decided that a swim would help her relax. I've looked right along the shore.'

'Let's hope she hasn't gone swimming, anyway,' said Jim. 'That's the thing, though. If we can't see her she must have gone somewhere on purpose. There are places to hide but I can't see how she'd get lost.'

Ben gasped. 'What the hell is that?'

'What?' Catriona froze. 'Where?'

He stood up and pointed. Not, then, something in his soup. There was a yellow light down by the river, and as we watched it circled and began to sway up the slope towards us, raking the gleaming wet turf.

'Lord have mercy,' muttered Jim.

They all watched. I drank the last spoonful of MSG soup before it was cold as well as green and salty. The spectral presence, it was obvious to me, was in possession of a torch and was wearing pants which were not a medieval fashion.

'Nina,' said Catriona. 'Her hair's wet.'

Her hair was dripping, but she was wearing her waterproof coat and her clothes were dry. She was looking down and her face was hidden by her hair. Bare feet poked from under her flapping jeans.

'Where have you been?' asked Yianni. 'We looked for you.'

She didn't look up. 'Please can I have some hot water? I need to wash.'

'Why? What have you been doing?'

'Please, Yianni. There are dead things on my skin. I need to wash them off.'

She rubbed her gloved hands. Drops of water fell from her hair onto her toes.

'Where've you been all afternoon?' He stood up and put his hand on her arm. She pulled it away and stood hugging herself. Did Lady Macbeth have OCD? Discuss.

'Nina, did you go in the river?' asked Catriona. 'You look cold.'

She raised her head for the first time, her skin yellow and eyes huge in the lantern light.

'I'm cold as ice,' she said. 'Please could I have a bath?'

Catriona and Yianni looked at each other. Yianni shrugged.

'There isn't a bath, Nina,' said Catriona. 'Do you remember? We're in Greenland. No plumbing.'

'But I'm so cold.' Her shoulders shook and she started to cry.

'Yianni,' Catriona put her arm round Nina, and was not pushed away. 'Heat some water. Come on, Nina, let's get you to bed. Yianni will bring us some water in a minute.'

Catriona led her away. Yianni went over to the stores tent and poured a few pints of our drinking water into the biggest pan. Lifting from the knees, he carried it back, re-lit the stove and balanced the pan. We watched until steam began to rise.

'She's not right,' said Ben. 'Now what do we do?'

Yianni ran his finger around the edge of the pan. 'Don't say that. Maybe she'll be better in the morning. She hasn't been sleeping.'

'And you think she'll sleep tonight? Seriously, Yianni, are there any sedatives in the first aid kit?'

'No,' said Yianni. 'And we're not going to try to drug anyone. If she's having serious problems we'll call the coast-guard and get her helicoptered out.' He dipped his finger in the water. 'I'll take this over.'

Ben and Jim and I looked at each other.

'Looks like serious problems to me,' said Ben.

'Not great.' Jim picked up the plates, where cold noodles lay like drowned worms after rain.

'Chocolate pudding?' I asked.

I dreamt one of the 'trauma dreams' that night, the one where I run up flight after flight of stairs because at the top James is trapped by flames. The stairs are slippery and I keep passing the bodies of fallen children, crumpled and bleeding on the landings. Sometimes I am carrying a baby, its warm head rolling heavily on my shoulder, and I bend to smell its vanilla hair and it falls, a single thud on the concrete, and is still, and James burns. I wake up, and I'm right. James burns. James burns, as if I'm keeping him in hell.

So I was already awake when the zipper on a tent ripped up and feet stumbled over the tent pegs. I lay there, wanting our own bed to cry in and thinking that returning to Plum Street could well be worse than it was getting into the cab and leaving his clothes and wine glasses to wait alone. There was rustling from one of the tents, and another zipper.

'Nina?' Jim calling. 'Nina?'

More footsteps, steady and fast. I sat up. A torch shone out.

'Jim. I heard something. Is she gone?'

'Her tent's open. Nina?'

'I'm here. Did you hear him?'

'No,' said Jim. 'I heard you. You OK?'

'Yes,' she said. 'But I don't like it when they come in the night. The dead ones. He opened my tent and looked in. Blood on his face.'

'Nina. Come on. Don't talk like that.'

'Did you hear him?'

'No,' said Jim.

'Ruth,' said Nina. 'You're awake. You heard. You were awake before he came. I heard you moving.'

'I wasn't.' I lay down again. 'I woke when you opened your tent and started stumbling about.'

'I haven't left my tent. It wasn't me. You heard him. The one they can't find.'

'Who can't find?' asked Jim.

I pulled my lavender pillow into the hood of my sleeping bag and pushed my hair back so it wouldn't tangle while I slept.

'I wouldn't fuel the delusion, Jim. If you take it seriously you reinforce it. Good night.'

'Nina? You going to be OK for the rest of the night?'

'It depends,' she said, 'which of them is about.'

'Good night, Nina.' I heard Jim zip his tent again. 'Sleep well. Don't get cold.'

'Nina?' said Yianni. 'Look, if it's bad, call me, OK? Don't just lie there feeling scared. I'm right here.'

I got up at first light. Yianni was up, smelling like old fish in the pants and sweater he's worn for days and hunched at the laptop again.

'You want breakfast?' he asked.

'Later. I want to get started.' The sky was white not grey. 'If the weather's OK I should get the first one out by nightfall. Save putting the canopies up.'

'Sure,' he said. 'Go for it. I'll come up in a bit. See how you're doing.'

'You can't have much power left there.'

The computer was cradled in his lap.

'Half an hour. I can get these notes done. The sun's nearly up.'

'I hope you're backing up.'

The turf was slippery with dew, and I transferred my tools to one hand to use the other for balance. At the edge of the pit I paused. The tarps were down, but not, I was sure, quite as I had left them. I'd used a lot of stones, edging in lines, paying unnecessary attention to the arrangement. Whoever had been there in the night had done a much more basic job. Some of the spare stones were arranged in a little cairn above the corner where I'd uncovered the long bone.

I left the tools there and went carefully back down the slope. Yianni looked up. 'Found something already?'

'Nina's been up there,' I said.

'Nina? In the burial? It's the last place she'd go.'

'I bet that's where she was. When we couldn't find her. Someone's been playing with the tarpaulin.'

He put the laptop on the groundsheet. 'You're sure?'

'Yeah. I used lots of stones as weights. Someone's built a cairn with them.'

'The burial's still protected?'

'It is now. I don't know what's been exposed. Yianni, if she's going to mess with the site we really do need to send her home. She could cause real damage here.'

'You don't know it was her.' He shut the computer down.

'Come on,' I said. 'Who else? We can't have her sabotaging the dig, Yianni. Think what the research council would say.'

'Let's have a look.'

We climbed back up. The sun came over the mountains and faint shadows appeared.

'Look,' I said. 'I certainly didn't build that. It's just where I want to start work.'

Yianni squatted down at the edge.

'Have they all been moved?'

'Don't know. I left a lot more than are here now.'

'Hm.' He looked up at the mountain. 'Nina does build cairns. I've seen her do it on beaches.'

'She's made some on the shore here,' I said. 'Bigger than this. I just hope she hasn't messed with anything else.'

'Let's get those tarps up and see.'

I jumped down and took the cairn apart while Yianni piled the other stones on the grass, and then we lifted and folded the blue plastic.

'Looks OK to me,' said Yianni. 'Anything look different?'

I shook my head, and went over to the corner. The soil was patted down as I'd left it.

'No. I don't think she's tampered with anything under the tarps. But Yianni, that's not to say she won't.'

'Leave it with me.' He stood up and looked down the hill. 'Jim's up. And Catriona. Breakfast in fifteen minutes?'

I promised Mom I'd stop skipping meals and mostly I have, but once I'd teased the soil out of the skeleton's elbow I wanted to go on, to see the shoulder and the wrist and how they'd left his hands. I have learnt – it's one of the few things I've learnt – that once you've accepted that you're going to go on living it is necessary to eat at stated intervals. But I wanted to see his hands.

James's hands were perfect. I used to watch them when he was cooking or fiddling with something the way he did. If you look, most people's fingers are a little warped, bent or scarred by whatever they do. My right index finger curves in, I guess from years of writing with a heavy fountain pen the way French schools like you to do. My thumb is scarred from the first and last time Papa let me open my own oyster, and recently my nails have ridged, presumably from some deficiency resulting from grief or Greenland. I'm keeping them polished but it still shows. You have that white line across your middle and forefingers, I guess from a cooking or carpentry accident? James's hands looked as if they'd done nothing but grow in the sun, as if he'd never tried basketball or barbecuing or mending a bicycle. His fingers were straight, not knobbly at the joints, and on the backs of his hands fine blond hair shone in the sun. He used to stroke my face with his cool fingertips, brushing over my eyes closed in readiness. They would have been crushed, those hands, between the crumpling dashboard and the steering wheel. My guess is his hands never even made it out of the wreck. And maybe he was still there, still alive.

I took a brush and began to smooth the black loam, the particle ghosts of muscle and blood, away from the wrist bones. The arm lay across the body and the rib cage began to emerge between the fingers. When I first met James you could see his ribs, even through those close-fitting T-shirts he wore. I liked his thinness, his quickness, as if there were nothing hidden, but I also liked it when he started working out and muscles rose across his torso. No one really needs a six-pack, but it was probably my favourite of the unnecessary things he acquired at Chase Garmon. I gave the bikes to his brother. The juicer and the bread-maker and the pasta machine are still there. I can neither use them nor give them away so they sit there as if they don't know he's gone.

The tips of the fingers were folded over the other hand. I knew that I'd have to lift both hands, and that the joints might come apart and lie like marbles in a box, but I left them shapely while I exposed the other arm. The bones were pale brown, the colour of the lattes I used to sip in Barnes and Noble.

'Ruth? I brought you some breakfast.'

Yianni stood at the grave-side, on my horizon.

'Thanks. I thought I'd skip it for now.'

'You're not ill? You need to eat.'

'No.' I knelt up on the soil. 'Just keen to keep going. Now I've started. You know what it's like.'

'Better to eat. You'll be digging all day. Look, it's not dried fruit. I'll leave it here. You've got wipes for your hands?'

He stooped and then squatted down and peered in. 'It's in good condition?'

'Yeah,' I said. 'As you'd expect.'

'Good.' He stood up again. 'I've got a couple of things to do – I want to chat to Nina, but I'll send Ben up when he's eaten and I'll be up later. Looks as if there's plenty to do here, huh?'

I looked at my skeleton, nestled in the corner, and the size of the pit. There was room for at least another four or five bodies.

'I'd guess so.'

'Remember to eat.'

He was silhouetted against the sky for a moment and vanished.

I turned back and began to free the collar bone. James had a hollow there that I used to kiss. His neck smelt of him, no sweat, none of the garlic and chilli that often clung to his fingers, nor the staleness of feet nor the sea smells below the belly. I used to bite his shoulders, sometimes, lying on his chest, and – sorry. Well, I suppose I already know I'm not going to send this to you. Though since we have to take all our trash back with us I can't discard it either.

I left the head for last, I don't know why. As if it were a gingerbread man or a jelly baby. As I'd expected, the pelvis was that of a man and he was tall, taller than James. I knelt in the mud and began to work on the feet. Feet are complicated, which makes sense when you think of something six feet tall and weighing one hundred and seventy pounds running and balancing on such little pads of flesh. I never liked James's feet. They always smelt and they were always cold, and he used to sit in front of the TV ferreting around in his socks and picking bits off his toenails until I wanted to take his shiny new cleaver to them. I always liked the idea of curling up together to watch a movie but in practice, on the rare occasions when we found something we could both tolerate, I had to sit where his feet weren't in my line of sight and keep knitting. I took a smaller brush to coax out the little bones in the toes.

'Hi there. You've done a lot.'

Ben loomed above me.

'I got up early.'

'And you were awake in the night.'

'Yeah.' I straightened my back. 'How's Nina now?'

'Asleep, apparently. Yianni's leaving her to it.'

'I don't see how we can have her digging. Did he tell you she's been messing around up here?'

'He said the stones weighting the tarps were disturbed.' Ben walked around the edge, deciding where to start.

'And built into a cairn. Like the ones she's built on the shore.'

'Didn't they use to use cairns as grave markers?' he asked.

'Who's "they"? Some prehistoric British cultures had funerary cairns with burials inside.'

'But the Norse ones were mostly memorial, right?' He climbed down and began to scoop out the opposite corner.

'Some of them. Some are just way-markers. Don't tell me you believe there's a Norse revenant erecting memorial cairns in the night?'

'No. Obviously not. I was just wondering. I mean, some of what Nina says – well, never mind. OK if I start here? Yianni said choose a corner and work out.'

'Be my guest.'

He climbed down and took up his trowel. He was wearing exactly the same clothes he'd had on for days, a pair of generic jeans and a dull red sweater under his waterproof parka. He looked up. 'Can I ask you something?'

I shrugged, thinking, no.

'Is there a reason you've done it like that?'

'Like what?'

'Leaving the head. It's like you're keeping the face covered.'

'Maybe I am.'

I came, at last, to the head. His hair burnt, all of it. I know because when I asked them for a lock of it, when they said I couldn't see the body, they said no. It's not so much the hair I mind; he himself cut his hair, hair doesn't carry pain. But his face. Think about it. Or don't, if you can refrain. Skin melts. And eyes – blue eyes that used to look at me as if I was some beautiful statue he'd just seen for the first time. He must have gone by then, surely. If I knew he was already gone I could stop forcing myself through those final minutes.

I worked around the skull with my fingers and a brush. It was dull and dry like wood left out in the rain. The teeth grinned at me through the soil before I cleared the eye-sockets, young strong teeth with plenty of eating and talking left in them. I was slightly daunted by James's teeth, and more so when I learned that he cared for them as if they were the family pet. No candy, even when I brought nougat from the Christmas market in Dijon. An arsenal of toothbrushes, several of which buzzed and jiggled. Three different kinds of floss which I borrowed indiscriminately when I remembered, and dental appointments with a frequency that made the hypochondriac French look British. (It's true about the Brits and teeth, you know. Nina should have had braces twenty years ago and if I had Catriona's teeth I'd get them capped.) They rewarded him by being white and straight as a picket fence, straighter and whiter than yours, Adam Blumfeld, until Earl Upton forgot that he was driving his truck as well as listening to the news.

I worked along the cheekbones. James was beginning to get tiny wrinkles below his eyes which sent him to the beauty counters for the first time in his life. He didn't believe me, but I meant it when I said I liked them. Maybe it's just growing up in France, but I admire the crinkles at men's eyes. There's something sexy about an experienced smile. I brushed the soil of the dead man's skin from his bones and wondered pointlessly what smile these grains of earth had been. You told me the dead live on as long as people remember them, that love keeps the dead alive, but that's not true. Love plus death equals nothing at all. Death kills, you know, that's the truth that puts you out of a job. There's no virtual James in my head. What lives on is my memory, which is part of me and not him. My memory cannot surprise me, call me in the middle of the afternoon with a explicit request for the evening, smile when I wake him with croissants on Sunday mornings. He is ash and bone, James. Gone.

My eyes filled as I began to dig deeper around the buried skull, curved like an eggshell. Ben in the corner was working intently with a brush and I knew he'd found another skeleton. I shifted around to hide my face and kept going, stroking the bone as if it were sore skin. His face emerged from the ground, and then at the back of the dome, just about where James's parting ended, the eggshell was broken, pushed in as if by a tentative cook. The edges hung together, polished where bone had been forced into brain. I cleared more soil and kept going, pausing to sketch the skull in situ before I turned it slightly and found a stone axe-blade cradled in the round head like an apple in a bowl, nestled in what had once been a brain, a brain that made ideas and words and, no doubt, surprise presents and dirty suggestions. A brain that was now black soil.

'Ruth? What's up?'

Ben squatted beside me, his hand on my shoulder. I looked up. Tears dripped onto my jacket. I had meant not to tell. I hid my face in my hands.

'Hey. What is it?'

I bit my lips together. 'I lost someone. I mean, he died. And I've been thinking about him.'

The bones lay inert, mocking.

'Someone close?'

'James,' I said. 'I've lost James.' When can I stop telling people?

The earth was soft between my fingers and my face.

'Ruth?'

'He was my boyfriend.'

'Oh Ruth.'

He reached out to pat my back. Ashes to ashes. I wanted to bury myself. His fingers moved over my back but I sat still, knowing I'd have to look up and let him see tears and earth on my face. I know you think I should have done more of this, wailing and rending, but the problem is that after a while you have to stop and nothing has changed. Histrionics only pass the time and then there is the awful moment when you have to stop and sit up and wash your face. I could tear my clothes and shave my head and he still wouldn't come back.

'This is recent?'

I still feel like a fraud, telling people the date. As if births and deaths should stop for more glamorous disasters. I pushed the tears off my face and sat straight, looking into the side of the pit, away from Ben.

'Last Fall. November twenty-first, actually.'

'Ruth, I'm so sorry.'

You're right, on this, I know. There's no hierarchy of loss with terrorism higher up than road traffic accidents. Comes he slow or comes he fast, death is the same for everyone. But it still feels stupid.

'Yeah. Well. He wasn't in the subway. He was driving. Not even going home for Thanksgiving. Just a meeting, an out of town client. And the guy coming the other way was listening to the radio and not looking where he was going. A big truck. Head on. He was dead on arrival. James. The truck driver is fine.'

'Listening to the news?'

'Yeah.'

There were roots of turf and grass sticking out of the side of the grave, and above us the clouds moved fast. I waited for Ben to say that it was an ironic way to die, and I felt my face smiling as if it knew a secret joke. There is, I know there is, something ridiculous. Collateral damage. Death by rolling news.

'Do you think he knew about the bombs? James?'

I tried to stop smiling and pushed my finger down the side of my boot, which was fastened too tightly and hurt.

'I often wonder. He usually had a CD on, in the car. Nick Cave or PJ Harvey. I like the idea that he died without knowing, except that it makes him feel even more gone. Part of a different city.'

I reached out to touch the skull, and years of training stopped me. Don't touch the finds unnecessarily.

'Ruth?'

'Yes.'

'What are you doing here?'

I pushed my hair back and soil fell in my eyes. 'Making a fool of myself. Sorry. I wasn't going to tell anyone. It's meant to be good to have new experiences. When you're ready.'

'I won't tell anyone else if you don't want me to. Though it's nothing to be ashamed of. Doesn't Yianni know?'

'I didn't tell him. My supervisor might have done. I didn't ask. Look, shall we get on with the digging? Yianni'll be up here any minute.'

I turned away from him and crawled back to the skeleton, now lying exposed to the grey light and fast wind. Bones are not meant for wind and sun and they crumble surprisingly fast without the layers of muscle, blood, fat and skin that make up people. I could have told Nina; bones are no good on their own.

'OK. If you want. If you're sure you're OK.'

'Yeah,' I said. 'I'm OK.' I began to lay out the rulers before taking more photos, knowing that the next stage might separate legs from hips and neck from skull as surely as an execution. You can get bones out intact but there's not much holding them together. 'And whether I'm OK or not, we need to get these guys out before nightfall.'

I looked at the skull. I needed to remove and sieve the soil from inside, and when I did I would be able to see the stone axe, the murder weapon, through the eyes.

'Had you been together long?'

Ben was at work again, tunnelling down somebody's rib cage.

'Two years. But it was the beginning.'

Tears prickled again. Two years can be a long time. Compared to his mother's loss, mine is easy. (I know. 'Ruth, you have the right to grieve. Only you know what you lost.') But she has lost her past and I have lost what was my future. Most of my future. Thank God for the doctoral thesis.

'Was he a grad student too? James?'

Yeah, I thought, a graduate student with clients and out of town meetings. I knew I shouldn't have told anyone about him, shouldn't let people who never knew him invent their own versions. I had decided not to. Stupid, stupid Ruth, to go weeping over the finds as if displays of hysteria make any difference. I can be so dumb. ('Ruth, you can't suppress these things entirely. You need to find ways of working through it.' Oh *piss off*.)

'Please, Ben, can we stop talking about this? This is a chance for me not to be the tragically bereaved one for a few weeks, OK? People at home have either lost patience and gone away or they're still bringing me soup and telling me how bad I feel and I just want not to be that person for a little while.'

Mark Twain says something about the person asking another to keep a secret expecting of his interlocutor a discretion of which he himself is incapable. I'd put myself in the hands of a small sweaty guy from a place they could probably bomb to bits without anyone on the international stage caring much at all, and I'd have to wait and see if he could keep it to himself.

'Of course. I see that. Only, Ruth, couldn't you be both people? Is it a choice between tragically bereaved and functional?'

'Yes.'

'OK. You know what you're doing. But I'm here, if you want to talk.'

You are, regrettably, there whether or not I wish to talk, I thought, you with your fat frog hands. The way people go on, you'd think the need to talk was equivalent to the need to breathe or eat. Talking makes no difference except to let people get off on vicarious bereavement. Or make a living out of it. I found last Fall that even taxi drivers like slowing down for a close-up of technicolor grief.

'So,' he said. 'How do you think these guys ended up here, then?'

It was already clear that his burial had not been as carefully interred as mine. The body lay on its side, arms out and legs curled as if asleep, and it was tall. Assuming that Ben had another young man, also with deliberate injuries, it was obvious how they got there, and until we knew whether that assumption had any basis there was no point in speculating.

'We'll have to wait and see, won't we? Depends what happened to your guy.'

I picked up the pad and sketched in the detail of the bones, still nestled where they had been left seven or eight centuries earlier.

'I hope Nina's OK,' he said. He was still looking at me, not working. I still, of course, had soil on my face.

'Yeah. Just let me concentrate on this a minute.'

'Of course. Yianni's a long time.' He brushed the ribs. 'Do you think she's still sleeping?'

I took a breath. 'How would I know? I've been up here since dawn.'

He looked up and me, and then went back to the ribcage. I took the photographs, and after a few minutes he picked up his trowel again.

I was still finicking around, postponing the moment when the bones might come apart, when Yianni reappeared. His hair, which he hadn't, unlike the other guys, cut short before leaving home, stood on end and his face behind the scrubby beard was paler than usual.

'He's out!' He peered at the shape laid below his feet.

'Yup. You have another boy. I've made drawings.'

'Great. And I guess we know the cause of death.'

'I'd say so.'

'That's a stone axe?'

'It is. I'll tell you something, though, it's not a stone from round here.'

He climbed down and squatted at the skull, like someone paying a hospital visit. 'No. Well, the Norse had iron axes, didn't they?'

'Mostly,' I said. 'We'll see, won't we?'

Yianni raised his eyebrows. 'Of course. Er, Ruth, you seem to have earth on your face.'

'Yeah. I rubbed my eyes.'

He looked at me for a moment. 'You be careful. I agree it looks like death from trauma but no risks, OK?'

Pathogens in the soil. It's not a good idea to rub decayed organic matter into your eyes.

'I know. I just forgot.'

Ben had stopped working again and was watching us. I

looked back at him. Yianni waited, puzzled. The wind sang through the grass above our heads.

Ben looked away. 'How's Nina?'

'Better than yesterday. Not so distressed.'

'So is she making sense now?'

Yianni shifted, looking at my skeleton again. 'Sort of. She's reading *Middlemarch*.'

'Still talking about taking a bath?'

'Not this morning.'

I handed Yianni the camera. 'Hold this while I climb up. I want some pictures from above. You know she's going to have to go home, don't you? She's not safe around the finds.'

He fiddled with the string. 'Let's give it a few days. It's not an emergency.'

'Yianni, she's out of her mind. She needs to go home.'

'We'll see.'

I looked down at them, two living men and two dead. I glanced behind me but there was no one there.

'We can't look after her here. And she's obsessed with the remains. She can't be trusted.'

'No. But the thing is, the insurance doesn't cover pre-existing mental illness. I don't want to discover after we've called out the plane that the department is going to have to find thousands of pounds when she's not even an archaeologist.'

He was looking at a point below my feet.

Ben looked up. 'Pre-existing?'

Yianni shuffled his feet. 'It was years ago. I told you. Before she met David. Jesus, you can't let one crisis shape someone's entire life.'

'You haven't told them, have you?' I said. 'You haven't told anyone she's here.'

'Oh, I have. I said she was a doctoral student from Oxford.'

'And you didn't say she does English. You didn't say you were using research funding to bring a friend along for the ride.'

He rubbed a circle in the earth with his foot. 'They didn't ask.'

I started taking pictures. There are pictures of James which they won't let me see, even though I keep telling them that I took forensic archaeology, I know what dead people look like. 'It's different,' said the family liaison officer. 'You don't want to see those pictures. It's different when it's your own. It has to be, or we'd all go crazy. Remember him the way he was.' So I keep track, in my mind, of the way he is, and in some ways it will be easier when we reach the end and there are clean bones to think about.

It was clear that night that Nina was not better. It was her turn to cook and usually she'd have been fussing with bits of plants she kept deciding were edible, nagging Yianni about ingredients he hadn't brought and then, I will admit, producing something more palatable than the pasta-pesto and watery soup with noodles that the rest of us dished up. When Ben and I came down at dusk, carrying my skeleton in a long box, the stove was still in its holder and Nina's torch glowed from her tent. We slid the box into the finds tent and I took the cleaning gel and went off, at last, to wash my hands and sponge my face. I touched up my face too, and when I came out Catriona was heating water.

'You're doing Nina's cooking?' I asked.

She pushed her hair back and looked up. 'She's not well. I don't mind.'

I sat on my preferred stone and stretched my legs out. Catriona watched the pan and the last birds flew through the dusk.

'You finished your burial?'

Burial doesn't finish. That's the point.

'One out. Ben's halfway through another. We've left him in situ with the tarps over.'

She looked round at Nina's tent, from which no sound came, and dropped her voice.

'Are there more?'

I spoke normally. 'Looks like it. At least another three or four, could be more.'

Nina rustled. 'I wish you wouldn't,' she said.

'Wouldn't what, Nina?' I asked.

'You're disturbing them. They won't like it.'

I looked at Catriona, who was stirring the water into a vortex. 'Nina, it's an archaeological dig. We came here to disturb them.'

'They won't like it.'

'Nina?' said Catriona. 'Do you want to help me cook? It's going to have to be pasta and pesto. We had noodles yesterday.'

The zip on Nina's tent went up and her torch beamed out, making the twilight suddenly darker. Her face appeared, uplit and hollow.

'Don't overcook the pasta. Use lots of water. There's nothing else to get wrong.'

'What shall we have for pudding?'

'Even with polenta I could have made something. You could probably make rice pudding with reconstituted milk. And cardamom. Really it needs rose water.'

'But mostly it needs rice, which we haven't got.' Catriona tipped a packet of pasta shells into the pan. 'There's some tinned sponge pudding. I could boil it with the pasta.'

Nina traced the path up the hill with the beam of her torch.

'Don't,' said Catriona. 'You're making me nervous.'

'They're not coming yet,' said Nina. 'Later. Sleep in the middle of your tent, Ruth. Harder to reach.'

Catriona's shudder shook the spoon against the pan.

'Where are the guys?' I asked.

'Yianni's having computer problems. In the big finds tent.'

'I knew he'd run out of power,' I said.

'It's the internet connection.' Catriona scooped a shell out onto the spoon, prodded it and put it back. 'Sites not updating, or something.'

'Don't put a tin in with the pasta,' said Nina. 'For one thing, we'll end up eating the glue from the label. Not to mention whatever's on the outside of the tin. And for another, you don't want to reduce the temperature suddenly when it's in the middle of cooking. It'll go gluey.'

A cry came from the shore. Catriona froze, staring at Nina.

'It's that woman,' said Nina. 'Looking for her sister.'

'It's a bird,' I said. 'A gull. That's all.'

'Bad enough,' said Nina. 'If it's got that virus.'

Catriona stirred the pasta, shoulders hunched. 'They've all gone to sleep,' she said. 'It's got too dark for birds now. I wish . . .'

'What?'

'Nothing. Not really.'

She lifted a shell again and watched as steam wavered in the blue light of the paraffin stove. Nina directed her torch down towards the shore, where it picked up waving grass and sent long shadows skittering across the field.

'It's a mistake, you know. You'll regret it.'

'What will we regret, Nina?' obliged Catriona. I inspected my nails.

'Bringing that man down here. He's lost. They look for him. They'll come here now.'

'Nina, for pity's sake,' I said. 'It's bones. We've all got them. If he was lost it was six hundred years ago and someone found him and buried him. Stop it.'

She blinked at me as if there was light in her eyes, which there wasn't.

'The pasta's done,' said Catriona. 'Ruth, I'm sorry, but I'm too nervous to go off up there and find the boys. Would you? And can you work the lantern? We'll need it to eat by, anyway.'

'Sure,' I said. 'You want me to light it before I go?'

She nodded, glancing round again at the hill and the path down to the sea. 'You'll be quick, won't you?'

Up at the finds tent, the boys were kneeling before the computer as if it were a little god.

'Dinner's ready,' I said.

'Yeah,' said Jim, staring at the screen.

I waited a moment. The screen lit their faces, picking them out in the dark.

'It's pretty dark in here. Dinner's ready.'

'OK,' said Ben.

It was like trying to get James away from a big game.

'We'll save some for you, shall we?'

'What?'

They weren't even pressing keys, just watching the screen as if it was about to give them instructions.

'Catriona and I are going to eat now.'

'What about Nina?' asked Yianni, without looking up.

'She's frightening Catriona,' I said. 'And dinner's ready.'

'Down in five,' said Jim. Yeah, right.

Nina was back in her tent.

'Does she want any?' I asked.

Catriona shook her head. 'She says the pesto's made with sunflower oil not olive and that potato flakes don't feature in Italian cuisine. And pesto should be served with spaghetti, farfalli are for a heavier sauce. I did try. She says she doesn't like eating the wrong stuff. We don't have spaghetti.'

'No. Well, seeing how Ben eats noodles I'd rather have farfalli.'

'Are the guys coming?'

'They're in the world of men and malfunctioning gadgets. I said we'd eat.'

She pushed the green-flecked pasta around the pan. 'I am hungry. And it's nastier cold.'

I handed her a plate. 'Eat.'

Nina was still silent and Catriona and I were discussing whether dried banana chips were better or worse than nothing for dessert when we heard the others coming down the path.

'Did you fix it?' I asked.

'No,' said Yianni. The wind flattened his hair. 'The sites won't update. The connection seems OK. Has anyone opened any attachments? Any weird e-mails?'

'No,' said Catriona. 'I haven't been on since Monday, anyway. Will you be able to sort it out? I know not much is changing but I'd like to know. About the epidemic.'

'Sure,' said Jim. He blew on his hands and rubbed them. 'The anti-virus stuff's all up to date. Might need to reconfigure the connection.'

'Just don't take any chances,' said Yianni.

'Will it be OK tomorrow?' asked Catriona. 'I wanted to check again.'

'Do my best,' said Jim. 'But we're not getting anywhere now.' He looked towards the pan. We'd put a plate on top in a gesture towards keeping it warm.

'Fucking machine,' said Yianni. 'Maybe I'll just have one more go.' He stared at his feet.

'Yeah,' said Jim. 'Let's eat, huh? I used to work on the helpdesk back home. Portability can be a disadvantage when you feel like chucking it across the room. We saw some strange accidental damage.'

'I just don't know what's wrong with it. Everything looks OK. But it won't bloody work.'

'I'll try again in the morning. Don't worry. Your data's safe, anyway.'

Yianni kicked at a withered plant.

'I can heat this up,' offered Catriona.

'Don't bother,' said Yianni. 'Don't bother.'

I didn't dream that night. I slept so deeply that it felt like coming up from the bottom of a well when I heard voices, and even then it was so dark that I couldn't tell if my eyes were open or not. I thought for a moment that James was there in the blackness, and put my hand out to my right, where he used to lie. When my fingers met five cold points on the other side on the canvas I woke up.

'James? James!'

Maintaining contact with one hand, I reached for my torch and pushed the end against the ground. There was a hand outside the tent, pressed to mine, and the outline of an arm shadowed on the canvas. It was too small. Nina.

'Who is it?' I called.

The hand slid down the tent, fingers clawing in.

'Nina?'

There was scuffling as something crawled along the canvas, brushing against my stack of folded clothes and knocking over my body lotion. The zipper on her tent went.

'Ruth? Is he still here?'

'Nina, stop this. It's not funny. We're tired, the rest of us. We work. It's too much, messing around in the middle of the night. However crazy you are, it's too much.'

She laughed. I didn't like it.

'Sleep well, then, Ruth. Sweet dreams.'

When I woke again, grey light was filtering through the tent and I could hear the stove purring. The air in my nose tingled with cold. I'd pulled up the hood of my bag sometime in the night, and lay there content as a swaddled baby.

'Any progress?' came Yianni's voice.

'Not yet,' said Jim. 'It's only been ten minutes. We should really let the battery charge, you know.'

'Just solve the problem, OK?'

I rolled onto my side, knowing I should unzip the bag and face the cold. And another day teasing bones from their resting places. I opened my mouth and blew slowly to see if my breath condensed, which it did. I folded my hands back into my chest and dozed again, pulled equally by the knowledge that morning had come and the certainty that I was warmer and more comfortable than I would be until night came back.

'Did you hear them, in the night?' Catriona. The light was stronger now and the stove had fallen silent. A spoon scraped a bowl.

'Ruth? Yeah, that's why I'm letting her sleep.'

Someone, probably Ben, took an audible slurp of coffee.

'How's Nina?'

'Still asleep. At least, not answering. She was probably awake a lot of the night.'

'Do you think it was her?' asked Catriona.

'What, crawling about? Who else?'

'I don't know.' There was a pause. 'I do wonder. You're sure she's asleep? She doesn't behave like that in the day. What if she's – well, right?'

Someone put a bowl down on a rock.

'Right?' Yianni said. 'You mean, what if we're being haunted by the people we're excavating but only Nina can see them?'

'I suppose so. Don't look at me like that. I'm just saying, she doesn't act like that when we can see her. She reads *Middlemarch* and talks about food.'

'Yeah,' said Yianni. 'Well, she's always done that. But she hasn't always claimed to meet dead people in the middle of the night. Or demanded baths when the nearest tap is about two hundred kilometres away.'

'Oh, nearer than that,' said Ben. 'There are farmhouses.'

'That's what I mean,' said Catriona. 'If she usually went round communing with spirits it would be obvious that she had problems. But she does think she's seeing them.'

'Well, don't you start thinking you are. There's a lot of work to be done, you know. Concentrate on that, OK?'

'OK. Just as you say. Shall I wake Ruth, then?'

'I'm awake,' I said. 'Thanks for the lie-in. I'll be with you in a minute.'

Since they'd already had breakfast and I knew Yianni was anxious to start I didn't bother with my face. Rainwater's meant to be good for your skin, if it's not polluted. My hands were drying out and I put a travel size hand-lotion in my pocket to use on site before I put my gloves on. Then I ran the brush through my hair, pulled the previous day's jumper over my clean T-shirt and crawled out into the day.

There was no sign of any further tampering on site. We lifted the tarps to find the other body as we'd left it, a relief figure in the soil lying like someone frozen in an energetic nightmare, with the legs twisted and arm flung out. James was an active sleeper, and he talked as well as apparently wrestling boa constrictors in dreams he couldn't remember when he woke. I used to listen, convinced that his unconscious was muttering the key to his psyche, or at least the key to why he wasn't asking me to marry him, but the only clear sentence I ever got was about repairing dress shoes. I even rooted through his closet the next day while he was at work, wondering how literally to take his unconscious, and they didn't need repairing. That was when I found the letter from Polly.

I never told you about that. You were just waiting, weren't you, all those sunny afternoons while the ice melted in the pitcher and the art deco Moulin Rouge coasters protected the coffee table, for me to reveal the fatal flaw in our relationship. Some reason why I wasn't really as upset as I thought I was, some curtain you could pull so I'd get back on the twelve step highway and get rolling. Well, this isn't it. They broke up three years ago, before I met him. So the letter was no betrayal, not technically, but he always said he left her, or at least told her, one Saturday morning when they hadn't exchanged a kind word for weeks, that she had to move out. She couldn't handle his job, kept nagging about how late he got home and how early he left, wouldn't sleep with him if he had to cancel dinner. Towards the end, slobbed around painting her nails in front of the TV because she said it wasn't worth putting on a show for an audience that only turned up one night in five. So I never complained, woke up obediently when he came in from the office at midnight, left meals on the counter for him. Learnt to keep myself pretty even when I was sure he wouldn't be home for hours. But he still had her letter. 'I have loved you so much and tried so hard to please you, but

I can't keep doing this. It's damaging me, waiting when you don't come and cooking what you don't eat and never seeing anyone in case you want me. I can't stop loving you, though. You know where I'll be.' I put it back where I found it, haven't thought of it since.

Anyway, Greenland. Low clouds were rolling over the sea and the light was menacingly yellow. Yianni stood looking for a moment.

'OK. I think we want all of you up here today. See if you can get this one out before that storm breaks. Even with a canopy, we don't want him lying in a puddle.'

Ben sat on the edge of the pit, pulling his gloves on. 'OK. You coming up here too?'

Yianni looked out at the sea again. 'Later. I've a few other things to see to.'

'What about Jim?' asked Catriona. 'Because this isn't the last burial in here.'

Yianni dug at the turf with his boot. 'I know that. He needs to get the connection working first.'

'What, the computer? We don't need internet access now, do we?' I asked.

He shrugged. 'I'll be up later, OK? Come and find me if you need me.'

Catriona put her gloves on and pushed herself off the edge of the pit. 'I hope it's not that there's news he doesn't want us to know.'

'What, the virus? You mean you think he's faking the connection going down?'

She turned towards me, mouth open. 'Of course not. I

mean, that hadn't even occurred to me. I just wonder why he's so anxious. Maybe he knows something we don't.'

'Don't go looking for trouble,' I said. 'There's plenty of it right here.'

She looked down at the form emerging from the soil. 'Yeah. But long-ago trouble.'

Ben hunkered down and began to tickle soil from the feet with a brush.

'And what we've brought,' he said.

I looked over at him. If he's going to keep dropping hints I'd almost rather tell and be done with it. One day, if I tell enough people, I might even believe it myself. James is dead.

'I'll start on the head, shall I?' I said.

I knelt down and began to stroke the earth where the hair must have been. Occasionally hair survives peat burials, but summer temperatures here are high enough to promote decomposition, and probably freezing and thawing rots flesh and bone faster than a more stable, higher temperature. It seems odd now to remember that the T-shirts I'm using as a second skin, even under pyjamas, were outerwear a few weeks ago. I've been wearing silk and flannel pyjamas from the first winter in Paris, when the company found us a magnificent Haussmann apartment in the *cinquième*, floor to ceiling windows, plasterwork spilling off a wedding cake ceiling, cute little fireplaces where we couldn't light fires and original nineteenth-century radiators. Papa loved it. I used to sweep across the wooden floors to breakfast wrapped in a quilt, and one day he came home and handed me a package from Galeries Lafayette, with perfectly folded paper and ribbons the way only French shop assistants can tie them. Under the layers of tissue were pale blue silk pyjamas, lined with the only flannelette toile de Jouy I've ever seen. And then six weeks later we were posted to Jakarta and since then I've had central heating good enough for little chemises. Mom and Papa have enough stuff to fill several houses in storage, waiting for Papa to retire so they can settle permanently in a *fermette* in Bordeaux (Papa) and/or a saltbox in Maine (Mom), but I've kept those pyjamas with me in Jakarta and then Saudi and in the UK, where I wore them a few times, and New York, and now they're here in Greenland, taking on the smell of tent-waterproofing and damp down.

'Hey,' said Ben. 'Look, sesamoids!'

Tiny bones in the feet that rarely survive excavation.

'The hands are pretty complete as well,' said Catriona. 'Though don't you lose less bone on dark soil?'

'Yes,' I said. 'And where it's not stony.'

'Shame you can't research what archaeologists miss,' said Catriona. She paused and looked down at the hand emerging from the soil. 'Do you think he was just dropped in here? Or she?'

'Either that or alive at burial.'

They both looked at me.

'OK. There's no real way of knowing. Prof. Mitchell said that once, about one in a messy position like this.'

Ben shivered. 'That's why I want to be cremated.'

'Yeah,' I said. 'Well, ask your family. The funeral's not for you.'

James had made a will. I guess bankers keep their affairs in order. He had left no indication about a funeral. That was all his mom's idea; she planned it like the wedding that never happened, only with herself as the bride. Walking up the aisle with all those flowers.

Catriona was looking at me. I took a breath. Maybe I can be both people at the same time. Here is a safe place to try it out, with people I'll never see again. A trial run.

'I had to plan a funeral last year. My boyfriend died. A truck hit his car.' The teeth were beginning to gleam through the soil, still in their sockets.

'And he was killed. Immediately.'

I took a paintbrush and began to brush the teeth. The mouth seemed to be wide open.

'And I was at home and I didn't know.'

I looked up at her. She sat quite still.

'We were together two years. He's been dead nine months.'

'Oh Ruth,' she said. 'I am sorry.'

'Sorry' is not, I eventually pointed out to James's mother, an apology, but an expression of regret. People are claiming to feel sorrow, not responsibility.

'Yes,' I said. 'Thank you.'

The incisors were still there, but the upper canines had fallen out. Catriona was still, watching.

'He was buried. I keep thinking how long it will take. Though he was very burnt. And I don't know the soil ph. His parents' church.'

'Did you not want that? Burial?'

The lower incisors were also still in place, under the yawn or scream.

'I don't know. He was already so burnt.' I put down the brush. 'I guess I didn't want to dispose of the body at all. I'd like a mortuary. At least an ossuary. Like those painted skulls in Hallstat. So I could see him instead of thinking about him all the time.'

Hallstat is a pretty town in Austria where, in the eighteenth and nineteenth centuries, the skulls of the dead were exhumed, painted with flowers and the name of the deceased, and preserved in grinning rows.

'Really? Have you seen them?'

I shook my head.

'Only pictures. I just want something. Part of him.'

I picked up the brush again. One lower canine was gone. Catriona went back to the intricate puzzle of fingers.

'When my granddad died, Mum said what she really wanted was to push him out to sea in a boat. Like the Vikings. He messed about in boats all his life.'

I shook my head. 'You'd spend your whole life waiting for the boat to come back. You'd have to set fire to it and watch. But those bodies can't have burnt, you know. Once the boat got down to the waterline the body would just sink, wouldn't it, and you'd still be thinking about fish eating hands that had touched you and hair in the seaweed. I want something I can see and touch.'

'That's a bit creepy,' said Ben.

'No,' I said. 'Everyone you've ever kissed will die one day. Like this. Every hand you've held will rot.'

Though the list, it occurred to me, was probably not long.

'Yeah,' he said. 'But that's almost a reason to hold hands,

isn't it? You don't need to think about death, it comes anyway.'

'Don't, then. But one day you'll have to.'

If you're lucky.

'Ruth?' asked Catriona. 'Are you OK?'

'Trying to be,' I said. 'Wondering what it would feel like.'

I'm still broken, aren't I? I guess I'm beginning to realise that I won't get over it. Death doesn't get better. Maybe life does. The jawbone jutted forward and I began to work back towards the skull.

Yianni came up before the rain started, just as we were beginning to measure and photograph.

'Any joy with the computer?' asked Catriona. She had mud on her face where she'd wiped her nose on her sleeve.

'No,' he said. 'Jim's been trying all morning. It's maddening, there's nothing wrong with it. But all the sites are two days old and we can't get onto e-mail.'

Catriona stood up. 'Can I try? Maybe Edinburgh's still working.'

'I suppose so. I suppose you'll all want to. Listen, are you sure you haven't opened any weird mails or attachments? Not been on any dodgy sites? I mean, I've checked the history, but can you think of anything at all?'

'We told you,' said Catriona. 'Yianni, we've got as much invested in this as you.'

'What about Nina?' I asked. 'Did you ask her?'

'She's only been on e-mail. It all looks OK. But then why the fuck won't it work?'

He still hadn't looked down.

'Yianni?' I said. 'What do you think?'

'What do you mean? Oh.' He slid down. 'That's interesting.'

'Yeah,' said Ben. 'More bone injuries.'

The left collar bone was severed, an injury which ran on across the ribcage and into the pelvis.

'So it was a battle?'

'With whom?' asked Catriona.

'Pirate raiders, other Norsemen, the Inuit,' I said. 'Unless you can think of anyone else who might have been around. Lost tribes of Israel.'

'And the survivors buried the dead?'

'Can't see the enemy doing it.'

'No friend buried this one. Or threw him in. And I don't see why pirate raiders or other Norsemen would be using stone axes.'

'No,' I said. 'But this wasn't done with a stone axe. Maybe he was on the other side.'

'We'll see,' said Yianni. 'Just get him out, OK? Careful, but quick. I really want to get these guys out before we have to leave. However many there are. I'll bring you some lunch. And this afternoon we'll all be up here.'

If asked, I would have said that I thought it was a bad idea to leave Nina unsupervised with all the finds.

When we came down at dusk, Nina's tent was dark.

'Perhaps she's asleep,' said Catriona.

'Have a look,' I said, opening my own tent. 'I'm going to wash and brush up. Who's cooking?'

'Jim.' Ben stood by his tent, gazing at the dark sea. 'Only

Yianni told him to get on with the computer instead. I'll do it.'

I looked round at Catriona. It's hard to mess up the food we make here, but we've both seen Ben pick his ears while cooking.

'I'll do it,' she said. 'You help Jim.'

It's easier to let men believe that machines and gadgets work better with more Y chromosomes in the immediate vicinity.

'If you're sure.' He scratched his head slowly, and then began to dig at his scalp with a fingernail, trying to dislodge something the rest of us didn't want to know about. I turned away, and then back.

'How come they've still got power? For the computer?'

'There's not much,' said Ben, still scratching. 'That's why Yianni's getting frustrated. Here they are.'

Jim and Yianni came down the path, carrying another long, coffin-like box containing the earthly remains of someone whose afterlife was about to relocate to a university lab. Most of James was beyond recycling, but he carried a donor card and his mom went on and on about his heart, which had come through just fine in its cage of bone. Apart from the small fact of it not beating anymore. She liked the idea of her baby's heart still beating to someone's tread, but I needed him whole, needed to be able to follow him to the point of dissolution. And then, I suppose, one day, to let go.

They slid the box into the tent, next to my guy with the stone in his head, and came down to us. Catriona knelt by the stores tent.

'Yianni? Can I ask something?'

'What?'

'Have you tried the satphone?'

He looked over her head, towards the river. 'Why do you ask?'

She pulled the tent zipper up and down, making a noise like tearing paper. 'I was thinking. If we haven't got the internet anymore. We can still make contact, can't we?'

'We'll get it working,' he said. 'Don't worry. The plane's booked for the fifth, anyway. I said we'd confirm but they know we're here, they'll come anyway.'

'How do you know that?' Ben's shoulders hunched. 'If you said we'd confirm?'

'They know we're here, OK? Come on, Jim. Let's get that machine going. I'll light the lamp.' He looked around. 'Where's Nina?'

Catriona began to pull things out of the stores, holding them up to her eyes in the dullness.

'Are you hoping they'll turn out not to be noodles?' I asked.

'Maybe they changed into couscous while we were away,' she said. 'Or cake mix. What else comes in little boxes?'

'Gauloises. Tampax. New make-up. Jewellery. If you could have one little box, what would it be?' I reached in for my hairbrush and stood there, working up from the tangled ends. Catriona shook some of the packets.

'Honestly, probably orange juice or cherry tomatoes. Really ripe ones, you know? Though I wouldn't turn down one of those heat packs you can put in boots or pockets to warm them up. You?'

I began to work down from my parting.

'A ring,' I said.

She put the box down.

'You mean a particular ring?'

'No. Just an idea. I mean, I could have brought a particular ring, couldn't I? Nina did.'

She took the stove out of its box. 'Does it help, being here? Did you think it would?'

'Maybe. I sometimes wonder if I'm doing this instead of therapy. I mean, I think I'm a practice-based person. Better to keep busy. It's no worse, here. I don't know what going home will be like.'

'Yeah. I can imagine. I hope – well – I'm worried. What do you think is wrong with the computer?'

The last streaks of light faded over the sea.

'I'll get the lamp,' I said. 'No idea. Maybe when testosterone has failed we'll be allowed to try. I wouldn't worry. Like Yianni says, we don't really need it, do we? As long as they come and get us at the end.'

'Mm. That's what's worrying me.'

'We'll be OK,' I said. 'I'm less sure about Nina.'

I went to fill the lamp. I think it seemed as if losing James gave me some immunity to further disaster, as if there's a quota for misfortune and once it's filled, it's full. No cancer for me, no more sudden loss, no house fires. Masonry won't fall on me, my building won't have legionnaires', my car won't crash. I won't get stranded on the west coast of Greenland with winter closing in and a psychotic British woman in the next tent.

Yianni reappeared.

'It still won't fucking work. There's nothing wrong with it, it just won't work. Fucking thing. Now it's run out of power.' He bit his lips and blew like a horse.

'Must be something wrong,' I said, dangerously.

He grabbed a handful of hair on each side of his head and pulled until the skin on his temples curved out.

'Of course there's something bloody wrong. We just can't see what the hell it is.'

Catriona poured water into a pan. 'Maybe it's not the computer,' she said. 'Maybe it's the websites. Have you tried anything hosted outside Europe and North America?'

'Oh, shut up,' said Yianni. He knelt by his tent, reached in for a torch, and set off towards the river. 'I'm looking for Nina,' he called. 'Go on and eat.'

We were examining some gelatinous brown foam concocted by Catriona and called, she claimed, 'angel delight', when we saw the light again.

'Do you think he's found her?' asked Catriona.

'If not, I don't know what we do.' Jim bounced his spoon on the brown foam, making an unpleasantly wet noise.

'Coastguard on the satphone, I suppose.' Ben lifted a spoonful, looked at it and turned it over. The angel delight peeled off the spoon and landed wetly back on the dish. 'Though there's not much they can do in the dark. I think my nan used to eat this. And tinned rice pudding.'

'My gran used to make it for us,' said Catriona. 'A secret treat. We only liked butterscotch, the others were too chemical. Though she used to make them all anyway, as if you had to eat the pink ones to get the butterscotch. Like seeing bad films as the price of good ones or snogging frogs to get to the princes. I'm a bit worried about the satphone. Yianni won't talk about it.'

'Butterscotch?' asked Jim.

'Like toffee.' She took a spoonful and swallowed. 'Is it always a mistake, nostalgic eating?'

The torchlight came closer. There was only one pair of legs moving behind it.

'She's not with him,' I said.

Catriona dropped her plate, angel delight down, and stood up.

'Yianni? I thought you'd find her?'

'I have. On the beach.'

'She's not –'

'She's OK. She says she has to stop them landing. I'd have had to pick her up and carry her back here.'

'Stop who landing?'

'Men with knives, apparently.' He switched his torch off and stood at the edge of the circle of light. 'I don't know what to do. She's not OK, is she?'

'Nope,' I said. 'You need to get her out, Yianni. Seriously. She's not safe and the finds aren't safe and if she's going on about knives it seems possible that we're not safe either.'

'Don't,' said Catriona. 'She's not like that. She's just scared.'

'People who are scared are dangerous.' I put my plate down. 'Really, Yianni. Call the coastguard. We could have her out of here at first light. She needs help.'

'Ruth, there's a difference between delusional and

psychopathic,' said Ben. 'Her ghosts aren't doing you any harm, are they?'

Yianni sat down. Catriona handed him a plate of cold noodles.

'It's just the insurance. We'd be liable. The department would be liable. I'm worried I'd never work again. It's only a few more days.'

'And she's sitting on the beach in the dark talking about men with knives. Come on, Yianni, she's sick.'

Catriona picked up the torch. 'I'm going to her,' she said. 'If she wants to wait for men with knives, I'll wait with her. We can't leave her there on her own. They're real for her, whatever we think. Imagine how you'd feel.'

She scrabbled about in Nina's tent and came out with a greying pale blue lambswool sweater, which she folded. Then she went into her own and emerged with a bar of marzipan chocolate.

'I was saving it for emergencies,' she said. 'And I think this counts. Did she have any lunch?'

'I don't think so,' said Yianni. 'I did offer. She just went on reading.'

'I'll take some water, then.' She filled a half-litre bottle from the big cask.

'Catriona?' said Ben. 'Don't get sucked in, OK? The only danger she poses is to your mind.'

Catriona wasn't listening. 'I'll try to get her back here for the night. If I can't, I'll move my tent down there. You can't just abandon her.'

The torch swung away over the grass.

'And then there were four,' I said. 'Shall we clear away? I'd like to get to bed early. It's been a long day and it's my guess we're in for a short night.'

I was right. I lay floating in warmth and the softness of my sleeping bag, wondering what it will be like to lie on a mattress near a radiator and hear traffic and see streetlights and an alarm clock. After a while rain began to patter on the canvas, masking the sounds of other people going to sleep. It's a different kind of night, I find, when you can't see what time it is, but I was still more or less aware of my thoughts when feet came stumbling behind a careering torch beam and someone breathed hard.

'Wake up! Quick! Yianni?'

It was Catriona, as expected. I presumed Nina was walking into the sea or fighting off her imaginary pirates. I adjusted the hood to keep my ears warm.

'I'm coming. What's happened?'

A zipper went.

'I'm still dressed. On the beach?' Jim. I can't think what he'd been doing, fully clothed in a dark tent alone. Praying, perhaps, although one could do that in sweatpants and a sleeping bag.

'There are lights. Someone's coming. Run!'

Catriona was recovering her breath. The sea reflects moonlight and starlight, which is why the most important navigation lights are coloured red and green. I pushed my hood back. 'What colour lights, Catriona?'

'Yellow. Like lanterns. Coming closer. Ruth, please come.'

'It'll be moonlight,' I said. 'There's no one out there.'

Another zipper went, and the beam from the big torch swept my tent, picking out the label on the body butter I bought from the Soins de Soi concession in Macy's two months ago.

'I'll go,' said Yianni. 'Ben, keep an eye on things here, OK?'

There was rustling from Ben's tent, on my left. 'OK. What things?'

'All of them.'

He was gone. The night settled again around me. I tucked myself away from the sides of the tent to keep dry, felt for James's cashmere scarf, folded under my pillow, and curled up. It stopped smelling of him months ago. He was wearing it the first time I saw him, wandering up and down Hemlow Street trying to find Ros and Hugo's party, and on nearly all our dates that first winter. He wore it like a Frenchman, twisted at the side with the ends flowing over his shoulders, and once when we sat on a bench having a conversation about whether to move in together and watching snow fall on Central Park, I braided all the tassels so tightly they stayed like that for weeks. Now it lives folded flat, though I have resisted my instinct to wrap it in a shroud of tissue paper.

'Ruth?' Ben called.

I turned over. 'What?'

'You awake?'

'I am now. Obviously.'

'Do you think we should look over the site?'

'No. What for?'

'Just to see. To confirm that there's nothing.'

'Go ahead. I'm staying here. And going to sleep. If any spectral pirates come, tell them not to wake me.'

Silence, and then more rustling.

'Jesus, Ruth. You're not rattled? Not at all?'

'What rattles me is Nina. She's not here. I'm going to sleep. Good night, Ben.'

He rustled about in his tent, sighing and coughing, but he didn't go up the hill, and after a while, with James's scarf pulled under my cheek, I went back to sleep.

I dreamt for the first time of Greenland. I went down to the beach and James was sitting on that rock, the one where the seals wait for whatever it is seals are waiting for. He was looking out to sea and I called and waved, James, James, they told me you were dead. I'm here, sweetheart, I'm coming. He didn't turn but I could see the sun on his hair and the curve of his back and I walked into the water, which sparkled like sun-warmed sea. It wasn't warm – even in dreams I am cold now – but I waded in, knowing that when I got to him he would turn round and hold me and death would have been a bad dream. I floated and swam and the waves cut off my view, but at the tops I could still see him, and as the troughs pulled me back I knew he was there. Spray stung my eyes and cold salt crashed up my nose. I coughed and kept going. I clung to the rock, scraping my hands, and reached towards him. The sea pulled me back but I held on. The returning wave banged my head on the rock. Before the next surge, I got both hands over the edge and pulled up, scratching my shoulder and grazing my chest. He was gone, traceless as a seal. The dreams used to end there and I'd wake newly bereft, without the habit of grief, but now they go on. I sat on the rock watching the waves. Light went out of the sky. I looked back to shore and there were no tents, no cairns on the beach, just ruins and empty sky. I sat there and nothing changed, and then I woke up and it was all the same. James crushed in a burning car, blood and muscle eaten by flames and still not dead.

'Wake up.'

'For God's sake, Ben, it's the middle of the night. Now what?'

'There's someone uphill. Crouching at the grave.'

'Tell her to go to bed then.'

'Come up.'

'No, I fucking well will not come. Deal with her, Ben. Good night.'

When I woke next it was light. I couldn't hear Ben snoring, nor Nina murmuring in her sleep. The stove wasn't roaring, spoons weren't clattering. A sheep bleated, nearer than usual. I sat up and opened the tent. The grass was stiff and white with frost. The sheep, behind Catriona's tent, looked at me. Droppings fell from under its tail and steamed on the turf. I grimaced at it and it loped away. The other tents were closed. The sky was pale with cold sun, and the waves streaked white.

'Hello?' I called. 'Good morning!'

Someone sighed.

'Hello? It's gone nine.'

Yianni yawned. 'What?'

'It's past nine. Where is everyone?'

'Oh fuck.'

He unzipped the tent and looked out, greasy hair over

encrusted eyes. I was glad I wasn't close enough to smell his breath.

'Fuck. We can't waste time. Wake up! It's morning!'

'Bad night, then?' I asked.

'Oh, God. Weird night. Poor bloody Nina. Come on, people. It's late. Get up, everyone.'

I looked at him. 'Weird?'

'There was a boat.'

'What, here?'

'Can we get up, please? We'll talk later, OK? Come on. Jim? Catriona?'

'I'm awake,' said Catriona. 'With you in a minute. Nina?'

'Hello,' said Nina.

'You OK?'

'Fine,' she said. Her tent opened. Her face was gaunt. The most effective ways of losing weight are not worth it.

'Hi, Nina,' said Yianni. 'You want breakfast?'

She shook her head. 'Not hungry.'

'You're getting really thin. What's David going to say?'

She shrugged. 'If I see him again.'

Catriona closed her eyes. 'You'll see him next week, Nina. Come on. You'd feel better if you ate.'

She shook her head and sat there like a snake peering out of a hole.

'Come on,' said Yianni. 'Get dressed. Ben? Jim?'

'Yeah,' said Jim. 'I'm coming.'

'OK,' said Ben. 'I'm fucking tired.'

I shut the outer flap and unzipped my bag. Cold bit immediately. I unbuttoned the pyjamas and tried to lift my top but the cold was paralysing. It was like trying to put a hand into boiling water. I pulled the top back down again and struggled into a clean turtleneck, and then added a wool sweater and cardigan. All those layers, no one knows if you're wearing a bra or not and one day isn't going to make anything sag. I put on clean inner socks and old wool socks before squirming out of the pyjama trousers and into leggings and jeans, but even through the leggings the jeans felt stiff and icy. I shuffled forwards, put on my boots and went off over the turf to brave the need to pee. I don't think Greenland is a place I will want to revisit.

Nina's tent stayed closed while the rest of us ate crackers, dried fruit from a packet and cold water for breakfast.

'So what's with this boat?' I asked. The water was so cold it hurt to swallow. I remembered the sour coffee in Barnes and Noble with forgiveness.

Yianni looked round at Nina's tent.

'Later,' he said, swallowing crackers. 'Let's get to work. I can't believe we've wasted so long.'

The waste, in my view, was spending the night ghost-busting on the beach. I took another mouthful of water and dusted the crumbs from my parka. I remembered something.

'Ben, did you wake me in the night? Something about someone at the pit?'

'No.' He didn't look up.

'No?'

'There was someone at the pit?' asked Yianni, cracker frozen in transit to his mouth.

'No,' said Ben.

'No?'

'No.'

'All right,' said Yianni. 'Come on. Let's get going.'

'OK,' I said. 'Back to the grave?'

Yianni glanced towards the pink tent again and frowned.

'Yeah. Careful, but quick. We're running out of time here.'

'I know the date, Yianni.' I stood up. 'You get anywhere with the computer?'

He and Jim exchanged glances.

'No.'

I could hear one of the tarps rattling and flapping in the wind before I got to the site. Yianni and Jim had not weighted them as carefully as I would have done, and one had blown loose in the night. As I reached the graveside, it came free and plastered itself to the side, banging against my legs. I pulled it up and fought until it was bundled up. Then I looked down to see the prints of bare feet where the skeletons had been.

'Hey!' I shouted. 'Yianni? Nina's been up here again.'

'What?'

'Nina's been trampling the burials. Get her up here!'

'What?'

I ran further down the hill. 'Come here!'

He came. 'What?'

'Footprints over the burials. Nina's been mucking about again.'

He looked at me. 'I don't think so, Ruth. She was on the beach with us.'

'Come and see. Someone's been down there.'

We climbed back up and looked down. The footprints circled the places where the bones had lain.

'Yeah,' he said. He spoke to the indentations. 'Someone's been here all right. But it's someone with bigger feet than Nina.'

I climbed down and set my foot next to one of the clearer prints. I take a size 9. Nina's a good three inches shorter than I am, and these bare feet were longer than my boots.

'Ben? He said he'd been up here, I'm sure he did.'

Yianni still didn't look at my face. 'When?'

'In the night. Full darkness. I'd been asleep. I wasn't really awake. He said something about someone at the burial site.'

'I'll go talk to him,' he said. 'But Ruth? Don't tell Catriona. She's getting a little freaked out.'

I stood up and went over to him. 'Tell me what happened last night.'

He sighed, still looking at the prints, and then met my eyes. 'All right.'

He sat down on the edge of the pit. I stood in front of him.

'Tell me.'

He began to kick his feet, and then stopped. He rubbed dead grass between his fingers.

'We got down there. Nina was standing on a rock, pointing. Very wet, by then. And Ruth, they were right. There were lights.'

'Moonlight.'

'In that rain?'

'Fishing boats. Secret NATO stealth ships. Russian nuclear submarines. Jesus, Yianni, you don't need supernatural explanations for ships in the night.'

'Then a boat came.'

'What kind of boat?'

He looked up again.

'An empty one. Wooden. Maybe two metres long.'

'Came where?'

'Oh, just drifting. Someone lost a dinghy, probably. But Nina started screaming. She said there were cowled men with no faces in it. Coming for the dead. Catriona got so scared she – she wet herself. She thought we didn't see, in the rain, but we did.'

'I take it there were no faceless men?'

Yianni dropped the crushed grass into the grave. 'No. But there was a boat.'

'Hm. Where is it now?'

He shrugged. 'Go look. We left it there. No one wanted to touch it.'

He stood up. 'I'll go ask Ben. What he saw.'

He shambled away, avoiding my gaze. Dear Lord, they all believe it. I'm stuck for the next week in West Greenland with a bunch of people in the grip of group hysteria, and we need to finish excavating a mass grave. What have I done? I waited a moment, looking down towards the beach, half-hoping to see the boat so I could join in. There was nothing, of course. I picked up my trowel and began to work on the next burial, keeping well away from the footprints.

I began to dig around the string outline of soil changes, which suggested that the person was tall and lying straight. I couldn't help noticing, close up, that there were a couple of handprints as well, big open hands with the palms pressed down hard. I faced in towards the corner and went on digging.

'Hey, Ruth.'

I looked up. Jim, seen from below, looks preternaturally tall.

'Yianni says someone's been up here?'

'Look.' I pointed to the prints. He whistled.

'Nina didn't do that.'

'No,' I said. I knelt down again. 'But somebody did.'

He climbed in and squatted. 'And the hands.' He held his own over them. 'Same size as mine, give or take. I suppose that exonerates you girls.'

I looked at his hands and the prints. They were about the same size. I looked at his boots.

'It's not really a matter of exoneration, I guess. As long as there's no damage. We've all been in here, if someone wants to walk around barefoot at night it's not wrong.'

God knows I've done stranger things. He was still holding his hand over the print.

'It wasn't me, Ruth.'

'OK.'

I went on digging.

'You want to see my feet?'

On a list of things I want to see, Jim's feet would be low.

'No,' I said. 'Are you working up here?'

'Why would I want to come up here and walk around barefoot? Anyway, it was raining, remember? I wouldn't expose the site.'

'Good,' I said. 'Though the rain stopped, sometime in the night. Or I guess the footprints wouldn't be so clear.'

He put his hand down and stared at the footprints.

'I hadn't thought of that. Rain hasn't fallen on these, has it? Where were the tarps?'

'One blew off as I came up. Can't have been very well weighted. By whoever it was.'

'Me and Yianni, yesterday. But someone else since.'

I shifted round and began to work along the further side.

'Yianni or Ben, since you say it's not you.'

'Yeah. Well, let's not get personal, huh? No harm done.'

'Sure,' I said. 'I don't mind. You going to get your tools?'

He climbed out and then looked back.

'What?' I asked.

'Nothing. Back in a minute.'

I was coming back round to where I'd started, leaving a coffin-shaped channel along one side of the pit, when Ben appeared. He, at least, was carrying his kit.

'Seems you were right,' I said.

'About what?'

'When you said you'd seen someone. What made you look, by the way?'

'I didn't say I'd seen someone. You were dreaming.'

'Oh, I dream, all right. But not about you. You woke me up and said you'd been up here and seen someone crouching by the pit.'

'I didn't, Ruth. I slept from when we talked until you woke me this morning. If you thought someone was talking to you in the night it wasn't me.'

I looked at him. He looked back, eyes wide, shoulders held high. I looked at his feet, and at the prints. He's a small guy, no taller than me, but his hands are big.

'You think I came up here in the night and took my shoes off and walked on the graves? Ruth, I'm not mad.'

'Someone is,' I said. 'Well, two someones. Nina and someone with bigger feet than Nina.'

'Yeah.' He turned and looked down towards the sea. 'Either that or there's another possibility. There's someone else here. Someone we don't know about.'

Cold slithered down my back. 'That's ridiculous. How could anyone be here without us knowing? It's open hillside. Not to mention there's no food.'

'There's food in the stores tent. We all think we've heard someone moving around at night. There are big rocks, anyway. And none of us have even looked in the next valley – there could be troglodytes in caves there for all we know.'

'Troglodyte caves that feature on no maps with no fires and no animals and nothing coming or going. Anyway, has any food gone missing?'

He shrugged. 'We haven't been counting, have we? Maybe we will now.'

'Maybe we'll just decide there's no reason people can't walk about barefoot if they want to. As long as they don't damage the finds.'

He dropped his tools and vaulted down.

'It wasn't me, Ruth. And I wasn't talking to you in the night.'

Ben and I had started the brushwork by the time the others appeared.

'Sorry I'm late,' said Catriona. 'Nina's up. She says she'll make lunch.'

She didn't look like someone who'd wet herself with fear the night before.

'You're chipper,' said Jim.

'Nina seems better. I think she's relieved we've all seen the boat.'

'I haven't,' I said. 'And I gather no one else saw the faceless pirates?'

They all looked at each other.

'The boat was weird,' said Jim.

'And has it gone now?'

Yianni bowed his head. 'Drifted away, I guess. The tide's out,' he said.

'Yes,' I said. 'Well. Shall we get on with the dig, then? At least this light should charge up the computer.'

But we had barely allocated tasks when dogs began to bark.

'What the –?' said Ben.

'Dogs, I should say,' I said.

We looked at each other, dropped our tools and climbed out. Two men on horses came along the shore, with three dogs slinking at their heels. The sheep began to swirl in panic.

'Bloody hell,' said Yianni.

Nina appeared from the stores tent – a thought about missing food flitted through my mind – took one look and crawled into her tent.

'Oh,' said Catriona. 'They're wearing jeans.'

'They're probably listening to iPods,' I said. 'Did you think they were Viking visitors?'

One of the men whistled and a dog streaked across the field to where a handful of sheep were running away from the bleating herd.

'Shepherds,' said Yianni. 'They must be taking in the sheep. I guess winter's coming.'

'Should we go down?' asked Jim. 'Seems kind of unfriendly to ignore them.'

Yianni looked around and sighed. 'Not everyone. There's too much to do. Anyone speak any Danish?'

'Nina speaks German,' said Catriona. 'But I bet even shepherds in Greenland have a bit of English.'

I thought about James, sitting up late watching travel shows about Amazonian tribes and Himalayan villages and then flying to Tokyo to spend three days at a bank and go straight back to the office from the airport. He'd have loved real Greenlandic shepherds on horseback.

'I'll go,' I said. 'I used to get by in Dutch. When we were in Amsterdam. But they will speak English.'

'Me too,' said Jim. 'I want to meet them.'

'OK,' said Yianni. 'Give them coffee. Ben, Catriona, you OK to stay here?'

'I suppose so,' said Ben. 'We've started so late, better get something done.'

'Just tell me if they know any news,' said Catriona. 'Anything about the plague.'

Jim and I went down the hill.

'Dutch?' he asked.

'I lived a few months in Amsterdam. You don't really need Dutch but it's better to pick up a bit.'

'I envy you. I lived in the same town my parents grew up in until I went to college.'

I kicked a stone, which rolled and bounced towards the tents, stopping against Nina's.

'I'd rather my kids had one home,' I said. My kids. Not James's kids.

Nina peered round her tent.

'Something –' she called.

'It was a stone,' I said. 'A loose stone. Can you get the kettle on? We're making coffee for the shepherds.'

She peered the other way. 'I'm scared of dogs.'

'The dogs are busy,' said Jim. 'They won't hurt you.'

'And if they really wanted to hurt you, a tent wouldn't offer much protection,' I pointed out.

Her shoulders emerged. The shepherds had stopped by the river while the dogs gathered the sheep into a corner. Jim waved to them and one of them raised a hand.

'Hello,' called the man.

Jim and I went down to them. They swung off their horses as we approached. Short men, Greenlanders, wearing jeans and knitted hats like Catriona's.

'Good morning,' said the other man. Dutch, I saw, was not required.

'Good morning,' said Jim. 'Can we offer you coffee?'

They bowed and nodded. 'Thank you. You are digging here? Digging the Vikings?'

'Yes,' said Jim. 'Just the farm and the chapel.'

'Ya,' said the first man. 'They farm like us, hey?'

'We think their sheep stayed out through the winter,' said Jim. 'But it was warmer then.'

'And maybe again. Short snows, last year. Global warming. Soon we grow fruit, hey? Apples and pears?'

He sounded as if Greenlandic farmers might be holding climate change food festivals in the near future.

'But you are not staying the winter?' asked the other man, rubbing his hands. 'In these camps?'

Jim shook his head. 'One more week. We have a lot to do. Would you like to come up? My colleague is making coffee.'

The men spoke to each other, a language apparently without consonants, and then attached the horses to each other and called to the dogs, which crept around marshalling sheep.

'A few moments. The sheep don't wait.'

We walked back up. Nina had the stove going and was moving around sideways, facing the dogs.

'Candy?' asked the smaller man, pulling a bag from his pocket.

I took one and unwrapped it. It looked and smelt like bubblegum. He offered the bag around and Nina declined. People like her shouldn't be allowed out of Islington, or wherever it is she lives. Her hands shook as she poured water into the granules.

'It's not good coffee,' she said, passing them each a cup.

They each took a mouthful and bowed again. 'Is very good. Thank you. Your digging is good?'

'We've found lots of bodies,' said Jim. 'Skeletons. Looks like there was a fight.'

'Vikings fighting?'

'Fighting someone,' said Jim. 'We don't know yet.'

'Vikings were good fighters.'

'Yes,' said Jim. 'Though they seem to have lost this one.'

We all sipped coffee. I wondered whether Greenlanders share the European liking for milk.

'You heard the news?' asked the taller shepherd.

'No,' said Jim. 'Our computer isn't working. What news?'

'There is the sickness.'

'What, here?' Nina spilt half her coffee over her boots. Plague comes to Greenland.

'No,' said the man. 'Not Greenland. But Denmark, now. England. America, of course. All Europe, now.'

Nina put her cup down. Her face looked grey.

'Is it bad, in London? Are lots of people dead?'

The man reached out and patted her shoulder.

'You have family there, I think. Perhaps not so bad. We hear the Parliament is closed? But not so bad. Just a – precaution. Not so many dead.'

Nina turned away and crawled back into her tent. We could see her rocking backwards and forwards inside it.

'She's not well,' I said. 'Please excuse her.'

The man frowned. 'There is bad news. It will be a strange going home for you.'

'It's a strange staying here,' I muttered.

Jim cleared his throat. 'And in America? In the Midwest?'

The men glanced at each other.

'There are dead, we hear. It is not good news. You are better here.'

They got to their feet and put their cups down.

'Thank you. Thank you for coffee. You need anything? We can take any messages? We have a computer at home, three days from here. E-mail. Not working last week, but my wife is mending it.'

'Not for me,' I said. 'Thank you.'

Jim hesitated. 'You could send an e-mail for me?'

'Of course.'

'Wait a minute.'

He pulled out his notebook, wrote and tore out the page.

'Tell my parents Jim is OK? And I can't wait to see them?'

The smaller man read the address. 'Please to write the message. Speaking English easier than writing.'

He took the sheet back and scribbled. 'Ruth? Not even to your family?'

'No,' I said. 'They know where I am. As our friend says, we're safe here.'

'You live in a village?' asked Jim.

'No. Just a farm. Like here. But with the computer.'

'Which isn't working?'

The man paused. 'No internet. This is not unusual. A few days, it comes back. A few days, it goes. My wife knows these machines.'

He glanced back towards his horses.

'Ours isn't connecting either,' said Jim.

'I will send your message. We must go now, we stay in a hut tonight. Come, Henrik.'

'Thank you,' said Jim. 'Goodbye!'

'Goodbye. I enjoyed meeting you,' I said. 'Nina?'

I sounded like my mother. She didn't reply.

Jim and I watched as the horses led the river of sheep away round the headland. The noise of bleating faded fast, and we were left with a new silence.

'I guess that's it,' said Jim. 'I wish I knew what was going on.'

'You will. Soon enough.'

We began to climb back up to site.

'We going to tell the others?'

'Yes,' I said. 'Secrets do groups no good. There isn't much to tell, anyway. He wasn't very specific.'

'Mmm. Odd that they've lost internet access as well.'

'You think the internet's caught the plague and died?'

'It's odd,' he said. 'That's all.'

We came to the pit. Catriona and Ben were working on the third burial while Yianni was making an outline of a fourth.

'Any news?' asked Catriona.

I looked down at her. Her hair, as always, was escaping from under her bobble hat, and her face was pink with cold. She wiped her nose on a glove that had been used for that purpose before.

'Not much,' I said. 'They say the epidemic's still spreading. I don't think they knew anything very detailed.'

'Their computer at home is down, too,' said Jim. 'I don't like the sound of it.'

Catriona looked up, mouth open. 'Where do they live? Not near here?'

'Three days away. On horseback. They're staying in huts while they collect the sheep. It's probably only about fifty miles.'

'What's wrong with their computer?' Yianni asked.

'He didn't really know. Something with the connection. Sounds like ours.'

'Oh.' Yianni put the string down on the ground. 'Could be coincidence?'

'Easily.' Jim climbed down and looked at Yianni's work. 'It probably is. I guess the chances of finding neighbours with computer glitches are high. It's just with the epidemic. You can't help but wonder if – well, if the sites aren't being maintained.'

Catriona dropped her trowel onto what sounded horribly like a skull. 'Sorry. What, that bad?'

Jim shrugged. 'We don't know, do we? And as far as I can see, we've got no way of finding out.'

In the silence, a bird's wings creaked overhead. I looked up to see a raven wheeling against the white sky.

'It's quiet now,' I said. 'Without the sheep.'

Wind sighed. The sea was rough and we could hear waves, a gull. Nothing else.

'Traffic and TV and computers are going to seem very loud, aren't they?'

'Let's hope so,' said Jim. 'Let's hope so.'

Digging went well the rest of that day. We got both skeletons out, one missing both hands, bones severed cleanly above the wrist with something very sharp indeed, one with the left side of the skull and jawbone sliced off and a cut mark in the collar bone where the blade had stopped. Quick deaths. Probably instant for the head trauma. I'd expected Nina to disappear again, probably followed by the discovery of some strange but inconclusive event in the valley, but at one o' clock she appeared suddenly on the edge of the grave. She looked down.

'I've heard those ones,' she said. 'Making a noise.'

'Is lunch ready?' I asked.

'If you want to call it lunch. I've put out food for you.'

She vanished again, but when we came down we found all the plates laid out on a blanket. She'd arranged concentric circles of crackers with what looked like pâté and pumpernickel with some kind of caviar, to be followed by tinned peaches and ginger biscuits.

'Pâté?' asked Catriona.

'I brought some odds and ends,' said Nina. If she'd had an apron she'd have been smoothing it.

'Pretty good odds and ends. Caviar?'

'It's only lumpfish. And I'm afraid it's black with food colouring. But it travels well. I thought it would make a change.'

She seemed to have changed channels, from horror films to cookery shows, overnight.

'Thank you,' said Jim. 'It looks great.'

We sat down and began to pass plates. Even Nina took a cracker.

'Nina?' said Yianni. 'Is this the last canned fruit?'

She froze and looked up. 'There's only a week left. I thought you'd like it. I was trying to help.'

'I do,' he said. 'It's fine. I just thought. I was keeping some back.'

We all looked at him.

'Just in case.'

'In case what?' I said.

'It's always good to have a reserve.'

'In case what, Yianni?'

'In case the plane doesn't come?' asked Catriona.

'I'm sure we'll be fine. Sorry. Don't worry about it.'

We ate in silence. I am looking forward to coming home, having a bathroom and a bed and not having to see any of these people ever again.

'Is there a problem, Yianni? Something we should know about?' I asked. 'Some reason you know why we should be saving food?'

'No,' he said. 'There's no problem. It's a good lunch, Nina. Thank you.'

Later, as dark fell, I excused myself and went down to the beach. I like beaches in the dark. That last break we took, the guesthouse in Maine, we used to stay on the beach every night watching the waves flash white in the darkness and the last birds calling. We held hands, walking over the pebbles, and sometimes clambered out to sit on the last rock where the sea surged at our feet. I kept hoping he'd ask me to marry him, sitting there between his legs with my back cradled and his arms around me. He never did. It would get so dark we couldn't see where to step and we'd slither back up the shingle and go for dinner at one of the seafood restaurants and then head back to the four-poster, which had been turned down with chocolates on the pillow. He never asked. He was never going to, was he? That's what you knew. He didn't want to marry me. He didn't want our babies. I was his present, not his future.

This evening I walked along the stones alone, on the other side of the sea. Moonlight moved jaggedly over the waves, and as I came to the corner of the bay I saw a shape tossing against the rocks. I teetered out to it and heard the thud of wood on stone and they were right. There is a boat, an old wooden one, rolled sideways, nudging at the shore. A dinghy adrift, lost from some dock, carried east on the current. I sat above it, watching the wooden seat tipping wildly. I wanted to climb into it and push off, but I guess the wind and current that carried it across the ocean and up the inlet have come to the end of the road and will bang the wooden sides on the rocks until the boat breaks up. After a while I came back up to the tents, and now I'm here again, sitting writing in a sleeping bag by torchlight, listening to the night and counting the days until I can come home. When I do – if I do – I think I might like to come and see you again. Will you have me back?

JIM

I've been postponing writing this, hoping it wouldn't be necessary. Maybe it's still a dramatic gesture, to write as if we're not coming back. To write, come to that, as if although we're not coming back someone's going to find this and send it to you. To write as if you're still there to read it.

Three days ago, we were getting ready to come home. The odd thing is I was kind of sad to be leaving. It has been so quiet since the sheep left. The birds had been massing on the water, shrieking and flapping like over-excited kids going on a trip, but flight after flight had left, low arrows sweeping the waves. The Norse would have been out on the water killing all they could, using these few weeks when there is still meat on the wing and the ground is cold enough to keep it. I thought about you, Dad. About you and Uncle Bill going out on the lakes, that time I went with you and realised that killing birds was never on the agenda. Those guns were just an excuse, weren't they, a ticket for a day spent rocking quietly on a little boat under the sky, away from Mom and Aunt Patty and all us kids. And unless my presence inhibited your heart-to-hearts, your outpourings of your inner lives in Deer Creek, you weren't even talking. Just sitting there watching birds while the guns slept under the lifejackets in the bottom of the boat. I never told the girls.

Anyway, when the quietness came here you could see what it was going to be like in winter. Every day there was more dark and longer cold. It seemed like the plants were withering and dying before our eyes as the birds and animals went away. It was like creation on rewind, back towards darkness and the void, and it seemed a shame not to stay and see it out. There must have been pleasure as well as fear, don't you think, as the time came to burrow into the house and tell stories while the landscape outside died and turned white? Rest, at last, from the bright nights and hard work of summer? It seemed like the real Arctic, the Arctic I wanted all along, was just beginning when we were going back to cities and rain and electric light. Sometime, I'd like to stay a winter up here, live through the dark weeks and see the promise of light slowly but surely realised. But not now, not with these people.

'But the Greenlanders would have taken electric light like a shot. Come on. Instead of rendering animal fat and trying to sew by firelight?' said Nina. 'You're romanticising.'

We were washing and labelling the last finds from the hall, a set of bone loom weights that meant the occupants had left in a hurry. Looms were big and intricate and necessary. You wouldn't leave one behind unless you really had to.

'Maybe,' I said. 'But they must have liked it here. At least at the beginning. They left the bright lights of Iceland. And Norway.'

She put down the pen – a freebie from Christie's auction

house – and blew on her hands. The fingers sticking out of her cut-off gloves were purple, the nails turning blue.

'You don't want to go home, do you?'

'Of course I want to go home,' I said. 'We always knew this would end. I'm worried about my family.'

I am. All the time. We have so little information here, a few words from a shepherd who doesn't speak English. I don't know if vast swathes of America are desolate, flapping doors, abandoned vehicles and rotting bodies, a disaster movie with all the special effects, or if the hospitals are busy and vulnerable groups are showing increased mortality. I don't know if it's you or me who needs to be worrying.

'Yeah.' She picked up the pen again and wrote out the label in her copybook italic writing. 'But you're not going back to your family, are you? You're going back to Boston.'

I thought about the house. She's right, you know. I forget that I'm not coming back to you. It is old and pretty but it's cold and nobody really keeps it clean. We all think the mess is someone else's responsibility. Mostly Harris's. And most of my crowd are abroad with fieldwork this year.

'Doesn't mean I'd rather spend the winter in a tent. I like it here, that's all. Doesn't mean I don't like being home as well.'

She shrugged. 'I can't wait. I'm only not counting hours because I can't multiply by twenty-four in my head. I have to do twenty-five and then I forget how many ones to subtract at the end. At least we'll know, then.'

See, Mom? An Oxford PhD and she's still counting on her fingers. I poured a few weights from hand to hand.

'I guess David's waiting for you.'

She looked away. 'Don't. I hope. I can't think about it.'

'Yeah, well. Me too. All of us.'

The long grass bowed and moaned and the water darkened under the wind.

'How long have you guys been together?' I asked.

'Four years.' She wrote another label.

'And you're planning the wedding?' She's asked me enough personal questions, and not always with sympathy.

She twisted her ring. 'Not exactly. I mean, I wish we could just have a civil partnership. I can't see how feminists can get married.'

'Most feminists were married,' I said. 'Virginia Woolf. Simone de Beauvoir. Wasn't Mary Wollstonecraft married to Shelley?'

'No,' she said. 'Students always think that.' She went on for a while about the family life of Mary Wollstonecraft and someone else. 'They got married because she'd been a single mother for years and it wasn't fun. Not because either of them approved in theory. De Beauvoir and Sartre weren't married. That was the point. And lots of feminist critics would say the Woolf marriage was an argument against.'

I took a course on modernism back in college.

'I thought he looked after her when she was mad and stopped her killing herself?'

Nina bagged the last of the weights. 'She managed in the end. Anyway, I'm not going to get married because Leonard Woolf nursed Virginia. Lots of husbands drive their wives mad and then don't look after them.'

I looked at her.

'What?'

'You sound better.'

She stood up. 'I'm OK. At least you all saw that boat. I mean, we might be haunted but I'm not mad. Or madder than the rest of you. I'm going to the loo.'

Loo is English for bathroom. Nina explained it to me at the beginning. Flush toilets began to spread in England in the nineteenth century when the battle of Waterloo was big news. Water-closet, Waterloo. Our nearest flush toilet is more than fifty miles away. She walked across the grass, shying suddenly as though something had peeked out from behind one of the rocks, looking around before she ducked into her tent for paper as if she were using an ATM somewhere risky. Her delusions seem so specific. It's weird, when I think about people hearing voices and seeing things that aren't there I've always assumed they'd be obviously insane, probably talking nonsense, smelly, acting like the more worrying tramps in Boston. Nina seems normal, at least in an eccentric British way, except that what the rest of us have been insisting are nightmares are clearly beginning to happen in the day as well.

Later, we were sitting around the stove after dinner. Ben and Nina had gone down to the river to fill the water-breakers for sterilising and the rest of us were huddled over cups of coffee. At seven o' clock, it was fully dark and too cold, really, to be outside. I thought about Mom dishing up and shouting up the stairs to Hannah and Holly.

'I just want to say, while she's not here,' I said. 'I think

Nina's seeing them in the day as well. She's acting weird. Moving away from things that aren't there.'

'Oh God,' said Yianni. 'Really? Did she say anything?'

'No,' I said. 'But she wouldn't, would she? She knows we think she's mad.'

Catriona blew on her coffee. 'She might not be. Maybe they are there and we're making some kind of unconscious adaptation not to acknowledge them.'

'Maybe,' said Ruth. 'But wouldn't you say that a group decision not to perceive hostile spectres in an isolated situation fits some useful definitions of sanity? I mean, we could all go mad if we let ourselves. Quite easily. But we don't.'

'I suppose. And I do think I know what I can see. I mean, you can't really start doubting your own sensory data.'

'No,' said Ruth. 'You'd go mad.'

There was a silence. Waves broke on the rocks.

'So is it better to tell Nina she's having delusions or better not to contradict?' I asked. 'Because as far as I can see, telling her she's deluded makes her feel mad, which makes her more unhappy, and not telling her reinforces the delusion, which presumably isn't good for her either.'

'I don't know,' said Yianni. He swirled his mug. 'It's only another two days. I just hope it all settles when she's back home.'

'Not our problem,' said Ruth.

There was another pause.

'Yianni?' Catriona ran her finger around her mug. 'Yianni, you do know that the plane is coming for sure? Even if we don't e-mail?'

'They'll come. They know we're here.'

'But the arrangement was that you'd e-mail. Was that just to confirm the time?'

'Something like. They'll come.'

He was looking down. I couldn't see his face, in the dark.

'What about the satphone?' I asked. 'Have you tried?'

We heard voices and a torch came wavering up from the river.

'The plane will come, OK? What we need to do is concentrate on finishing off and clearing up the site.'

'You're not planning to come back?' asked Catriona. 'Next year?'

He was still looking down. 'Who knows. About next year.'

The voices were closer.

'Anything might happen.'

'You know,' said Catriona. 'I think I really believe the world will end in my lifetime. I've always thought the evidence was there. I think I believe it now. Without having to convince myself.'

'People keep thinking that,' said Ruth. 'Think about all those cults that barricade themselves in and then have to blow themselves up when God doesn't do it for them.'

'I know,' said Catriona. 'I'm not planning to do anything about it. But someday someone's going to be right.'

'Right about what?' asked Ben. He dumped the water-carrier in the stores tent and started getting the sterilising tablets out of the box. 'Won't it be good to have water that doesn't taste of chlorine? And juice and soda?'

'Beer,' said Yianni. 'Cold beer. And glasses of wine with dinner.'

'Dinner, for that matter,' said Nina. 'Salads and roast vegetables and fruit. Ripe melons and oranges. Lychees.'

'Doesn't sound very local and seasonal,' said Catriona. 'Right, about the world ending. In our lifetimes. I was saying, I think I'm starting to believe it. Though I did when I was very little. I used to keep my Flower Fairies umbrella under my bed because I'd seen a pamphlet that said when the fifteen-minute warning went you were meant to do something with a golf umbrella.'

'A golf umbrella?' asked Ben.

'You were supposed to remove the nearest door from its hinges and prop it at a forty-five degree angle,' said Nina. 'I don't think it had to be a golf umbrella. Then you were meant to get under it, preferably with a supply of milk and water wrapped in kitchen foil, and await the arrival of the emergency services. You'd think if they thought kitchen foil could repel radiation they'd tell you to wrap yourself in it, not the milk. It was probably just meant to keep people busy removing doors and stop them rushing into the streets to panic or have sex or whatever Thatcher thought people might want to do in their last fifteen minutes. Though I don't know how everyone was meant to know, anyway. It's not as if they had air-raid sirens. By the way, there's ice by the river.'

'Already?' said Yianni. 'It's been a warm summer.'

'Look for yourself, then.'

'I didn't mean I don't believe you.'

'Why not?' she said. 'If I see imaginary Greenlanders, why not ice?'

'Were you scared about nuclear war, Nina? When you were little?' Catriona put her hand on Nina's arm.

'Too much to think about it,' said Nina. 'We had a copy of *When the Wind Blows*. It looked like a children's book. Like the Father Christmas one. I couldn't understand how the grown-ups could carry on going to work and cooking and putting us to bed when that was going to happen. Mum used to say that if things started to look bad she'd take us all to New Zealand, which she seemed to think would somehow escape nuclear holocaust. But it takes more than fifteen minutes.'

'My dad used to say we'd go and stay with the cousins in Skye,' said Catriona. 'But I thought the same thing. Being twenty miles off shore isn't much help in a nuclear holocaust.'

'I can't remember worrying,' said Ben. 'My dad probably reckoned that if it happened, it happened, and nowt he could do about it.'

I can't remember knowing anything about nuclear bombs until my teens. Is that right? I certainly don't remember being scared of anything except the Lavens' dogs and that Dad might die in a car crash like Will Johnson's dad. It sounded like things I've read about kids growing up in war zones.

'And you were ordinary kids in ordinary families? Your parents weren't radical activists or anything?'

Nina shrugged. 'Left-wing and politically aware. Like most academics. At least in the UK.'

'Not activists,' said Catriona. 'Mum kept talking about going to Greenham Common but she never did. I think it was more about being frustrated at home than feeling any urgent need to ban the bomb.'

'I think it was like that for a lot of women who did go. Wasn't the point that patriarchy was a wicked giant which imprisoned women at home with the kids with one hand and toted nuclear bombs around on motorways with the other?' said Nina. 'My mum took me for a few half-terms. The songs were good. But it's not very nice being six and knowing that you're someone's prison. And they wouldn't have my brother at all. Even if you're four, if you've got a penis you're the enemy. Men are irredeemable. Talk about biological determinism.'

'Greenham?' I asked. This British elision. Green'am. You know the 'h' is there but you don't say it. Green eggs and ham.

'Greenham Common,' said Nina. 'It was a protest camp. And a commune. Women only. Around an American military base. Mostly about nuclear weapons but, you can imagine, lots of other things as well.'

'I thought the UK welcomed the postwar American military presence?' said Ruth.

Nina looked at her. 'No one ever welcomes an American military presence.'

Ruth pushed her hair back. 'I'd say the people in the concentration camps probably did. Wouldn't you?'

Nina stood up. 'I'd say they might have welcomed it more earlier. I'm cold. I'm going to bed. Good night.'

'Sleep well,' said Catriona.

'Please,' added Ruth.

'Yes,' said Yianni. 'You should all get to bed. I've got some notes to write up.'

*

She didn't sleep well. She never does. When the nightmares started, I felt sorry for her. It was like when Hannah had her sleepwalking phase; it was fear like I'd never seen before and the fact that I couldn't see what justified it didn't seem relevant. But at least Hannah didn't come over all supercilious when the rest of us couldn't see the Dark Bears, and didn't expect us to tiptoe round them in the day. I guess after the three of us you know this better than I do, but after a while you get to resent the broken nights. Nina's not a child. We didn't sign up to be carers. And I know what you're thinking, you didn't raise me to turn away from the needy and the sick, I am my sister's keeper, but it's hard when the sick person is also given to unpleasantness. I was deep in warm sleep, dreaming that the five of us were gathered around the table. Grandma's red check cloth was there, before Holly spilt the ink, and Hannah was in her high chair, waving that Mickey Mouse spoon. Before the girls started to grow up. When you two would suddenly smile across the plates like you had a private joke. There was snow in the yard but we were warm, and then suddenly I was cold and my lower back hurt and someone was muttering outside in the dark.

I pressed my face into my sleeping bag and whispered through closed teeth. 'Oh, shut up and go to sleep.'

Then I sat up.

'Nina? You OK?'

'It's out there,' she said. 'Right by your tent.'

Then there was more muttering, sounding close. My scalp crinkled.

'Nina? That's you, right?'

'You mean you can hear him too?'

She sounded excited. I listened for a minute.

'No,' I said. 'I heard you. Get some sleep, OK?' I remembered what Mom used to say. 'Morning comes faster when you sleep.'

'I thought you heard it. I thought it wasn't just me.'

'Go to sleep,' I said.

I lay down again, and again heard the voice. I'm so reluctant to say this, but it didn't really sound like Nina, and it did, honestly, sound as if it were coming from between my tent and hers. I couldn't hear words, only a low chunter. Something – someone – reciting a grievance, maybe even a prayer. It could have been someone talking in their sleep, maybe Ruth or Ben. God knows Ruth has reason to – she lost her boyfriend in a road accident last year. It's too easy to get scared when there's no light and someone telling ghost stories in the next tent. I pulled my hood close round my ears and began my own whispering. I will lift up mine eyes unto the hills, from whence cometh my help. My help cometh from the Lord, which made heaven and earth. He will not suffer thy foot to be moved: he that keepeth thee will not slumber. I'm all right here, OK? The words for this are banal, but I think I'm beginning to understand some of the things I've always heard. If we are all held in God's love, then maybe in the end it doesn't matter all that much if we meet again in Deer Creek or in the sweet hereafter. I feel as if I'm already on the way, here, so far away. I know the point is that, if love survives death, the worst is not terrible. Painful, but not terrible. If. When it feels as if the end might be soon, it can be hard to hold onto that moment of certainty.

JIM

I woke to the purr of the paraffin stove and the rustling of bags. There was grey light, not enough to see colour in my tent, but enough to get up. Enough to start the last day here. I opened the flaps. Yianni was tipping dried fruit into the pan, his breath hanging in the half-light, and Catriona was sitting hunched in the entrance to her tent, fully dressed but silent. She wears a striped knitted hat with earflaps and fancy strings that hang down like Native American braids.

'What's up?' I asked.

Yianni glanced up. Catriona rested her forehead on her knees.

'What?'

'I made Yianni try the phone.' Her voice was muffled. 'There's no connection.'

'What?'

'We're completely cut off.' Her face lifted and she began to laugh as tears trickled down her cheeks. 'We can't make contact. We're stranded. We haven't even got much food left.'

She sniffed and put her head back down.

I looked at Yianni.

'Really?'

He shrugged, pointlessly stirring the fruit in the water. 'I'll try again. We'll have another go at the computer.' He looked up. 'We're not stranded, Catriona. They'll come. They know we're here. They'll come for us. Maybe they'll even come early when they don't get our call.'

'Or maybe they're all dead,' she said. 'Maybe there's no one left to come for us.'

Yianni looked towards Nina's tent, reached over and shook

Catriona's shoulder. He wasn't gentle. Her head swayed against her knees.

'Stop that. Shut up. I don't want Nina to hear this, OK? If you can't keep quiet, go away. We have to protect Nina.'

I reached in and unzipped my bag and came out in socks and the layers I wear to sleep. It was too cold. I crouched next to Catriona and put my arm round her.

'Nina's not the only person on this dig,' I said to Yianni. 'Other people get to cry as well. Hey, Catriona. Come on. We'll be OK. Maybe the phone never worked, have you thought of that? We never tried it before, did we? Even if lots of people are sick, the satellite wouldn't fall out of the sky. Satellites don't get viruses. Come on, we'll get everything ready, and tomorrow that plane will come. In forty-eight hours you're going to be home with your family, eating whatever you want, taking a hot bath, calling your friends. We're all going to be OK.'

She put her head on my shoulder.

'I'm so scared. What if this is it?'

'This isn't it. This is a phone that didn't work. It's not the end of the world.'

'But it might be. We've no way of knowing. Things wouldn't look any different. If everyone's died.'

I rubbed her shoulder. 'They wouldn't look any different if your Mom's making up your bed and getting the ingredients for your favourite meal and the pilot is thinking he hasn't got our call and maybe he'll come out first thing just to check up.'

Yianni took the pan off the stove and turned the valve. Silence, again. Nina's tent opened and she looked out.

'You don't have to protect me. I'm not stupid. I heard you trying the phone.'

She came out. We've all lost some weight but she is really thin now, which is odd considering how much time she spends in her tent reading while the rest of us dig.

'There's nothing we can do, you know,' she said. 'I don't think Jim's right about it all being OK but we might as well pretend he is. Think about it, Cat, there's going to be plenty of time to panic later.' She turned to me. 'Jim, you heard him, didn't you? In the night? He woke you up.'

I looked at her, so gaunt and eager. Even her hair looks thinner and duller than it did. If. If things don't go well. Well, I'll just say Nina's got nothing to spare.

'I thought I heard a voice,' I said. 'But Nina, it was the middle of the night. People talk in their sleep. Come on, we'll be out of here tomorrow. You can get some help, huh?'

'I don't need help,' she said. 'I just need to go home. If you think people who communicate with invisible presences and see the dead rise and walk the earth need help, you're ahead of me in the queue.'

I breathed in slowly and counted five as I breathed out. She's sick and she's scared and she has no experience of faith.

'Come on,' said Yianni. 'Let's eat. There's a lot to do today. Ben? Ruth?'

'Coming,' said Ben.

'With you in a minute,' said Ruth. 'Nina, that was so uncalled for.'

'Enough,' said Yianni. 'Breakfast. We need to make the most of the light.'

He passed around plates. Ben and Ruth came and sat down, and I noticed that Ruth wasn't even wearing lipstick. She's always pretty, but she looked drawn and pale. I guess we all do. I haven't seen my own reflection since I used the men's room in the community centre when we picked up the horses, but my clothes are getting a little big. Our diet hasn't been optimal. I can spend a lot of time thinking about what I want to eat – hamburgers from the barbecue with ketchup and mustard and pickles and all the salt and spices that don't earn their weight here – but mostly I'd just be glad to see you all again.

'So,' said Ben. 'What are you all going to do on Saturday? I'm going to lie in, then phone all my friends and go out for a real American brunch. Pancakes, bacon, sausages, syrup. Juice. Real coffee. What've you got planned?'

I watched Catriona trying to push a prune through the hole in an apple ring with her spoon. I could have eaten what she didn't want, no problems.

'Saturday,' said Nina. 'We could try to catch some fish. I suppose there's no way of getting a goose.'

Yianni put his plate down and stood up. We all sat quite still as he took three steps towards Nina, grabbed her wrist in one hand, drew the other back behind his shoulder and brought it hard onto her face. The slap thudded across the damp air and she cried out. He looked down at her for a moment and then walked off up the hill. She put her hand to her face and looked round at us.

'Hey, Yianni –' I said. Hey what? Don't hit girls? Come back here?

'You deserved that,' Ruth told Nina. 'You were being deliberately provocative. I'll go.'

She got up and followed Yianni, who was already halfway to the hall.

'Don't start a fight,' Ben said to me.

Nina was still sitting there. Her eyes filled.

Catriona put her arm round her. 'I'm sorry. I was just so shocked. We should have done something.'

'Sorry,' I said. 'I was closest. I should have grabbed him. I just didn't think he would do that.'

I put my plate down, not so hungry after all. We're all under stress but you'd hope it would take longer to start the *Lord of the Flies* stuff. Nina still hadn't spoken. Catriona patted her.

'He'll come back and apologise,' she said. 'I suppose he's just so anxious.'

'We're all anxious,' said Ben. 'But Jesus. Are you OK? Do you want some cold water on that?'

Nina shook her head, pushed Catriona away and stumbled back into her tent. The zippers scratched closed and then there was silence.

'What's she doing?' I mouthed to Ben.

He shrugged and mimed someone opening a book.

Catriona sat down again. She tried to poke the end of her shoelace back through one of the eyelets in her boot but her hand was shaking too much.

'Catriona, it's OK,' I said. 'He lost his temper. It's bad but it doesn't affect our getting home. It's OK.'

She looked up. 'It's criminal, apart from anything else. It's assault.'

'I know,' I said. 'If Nina wants to take things further when we get back, she can. But for now, let's just get through the day.'

I sat down. The next thing to do, I guessed, was to pick up the plates. Someone was going to need to go after Yianni and Ruth and bring about some kind of reconciliation, at least between Yianni and Nina, failing which I guessed we'd have to keep Nina on some kind of watch until the plane came. She needs Yianni. He's been the only one she'll talk to, some days. And then we needed to make sure all the finds were properly packed and labelled, take down everything we didn't need overnight and pick over the site for every sign of our lives here. I wanted to go back to bed and sleep, or at least lie quietly on my own, until I heard the engine. My bones felt too heavy and my back still ached from sleeping on the ground.

'We should clear up,' said Catriona. 'I suppose no one wants any more breakfast.'

Ben was eating again.

'We can't throw food away,' he said. 'There isn't enough left. They'll be hungry way before lunch.'

Catriona turned her spoon over and over. 'How much food is left, exactly, do you know?'

I stood up. I knew I'd feel better once I got going. You forget, don't you, how tired you are, until you stop again.

'No,' I said. 'And for now, all we have to do is pack it, OK? No inventories. We're on that plane in the morning.'

She rolled the spoon back the other way. 'Can we have one more go at the computer? And the phone?'

'It's Yianni's computer,' I said. 'I don't know where the

phone is. I'm sure he'll try it again. Come on, let's at least put the stove away.'

'I'm so tired,' said Catriona, not moving.

'Me too,' said Ben. 'But he's right. We'll look really stupid if the plane comes and we didn't get ready because we thought the world had ended. Try explaining that to the ESRC.'

The Economics and Social Sciences Research Council, who are funding most of the dig. UK taxpayers' money for Greenlandic archaeology. I stood and picked up the stove, which was painfully cold to touch. Ben held out his hands and hauled Catriona to her feet. She went and stood outside Nina's tent. No sound, not even the rustle of a page or a sleeping bag. I shook my head at her and she shrugged. When the plates were arranged in the stores tent and everything else put away she went back.

'Nina? We're going up the hill. But we'll be down soon, OK? We'll bring Yianni.'

Silence.

'Nina? Are you all right in there?'

'Go away,' said Nina.

So we did.

'You probably can't actually kill yourself in a tent in complete silence, can you?' asked Catriona, as we walked not very quickly up the hill.

'Depends what you've got,' said Ben. 'I should think an overdose doesn't get noisy till later. If at all. It'd probably be hard to slash your wrists without a murmur. But I think she's reading. It's tomorrow I'd be keeping a suicide watch. If the plane doesn't come.'

'We ought to stop talking about it,' I said. 'Assume it *is* coming.'

'I was assuming,' said Ben. 'Look what happened.'

We found Ruth and Yianni making a final fingertip search of the hall floor. 'Hi,' said Ben. We stood there.

'Hi,' said Yianni. 'I'm sorry. I shouldn't have done that.'

'It's not us you need to apologise to,' said Catriona.

Yianni knelt up. 'Partly it is. It shouldn't have happened. I just snapped. Sorry. How's Nina?'

'She went into her tent,' I told him. 'And she's been completely silent ever since. You need to go down and talk to her.'

'I will.' He started stroking the ground again.

'Soon,' I said. 'We're worried about her.'

He rubbed earth through his fingers. The same movement as Mom making pastry. Ruth continued her brisk progress.

'So, what do you want us to do?' asked Ben.

Yianni allocated tasks – Catriona and Ben to make a final search of the chapel, me and Ruth to check and refill the grave, lunch, packing and then a final survey of the site.

'It's got to be pristine, remember. As if we were never here. Or the fines are terrible.'

'We know,' said Ben.

Then the three of us stood and watched while he got up and went down the hill to Nina.

'Do you think they'll be OK?' asked Catriona.

'I don't think he'll hit her again,' I said. 'I don't know if she'll talk to him. They've known each other a long time.

They'll have to sort it out. Go on, you go down to the chapel. I'll help Ruth finish here. He's right, you know, we'll struggle to finish by dusk.'

I began to crawl across the floor. Did you ever hit anyone, Dad? In anger? I remember you smacking Holly once, Mom, after she pulled a pan off the stove and it nearly hit Hannah, and I guess that was anger. I had to haul Harris away from a fight one time. I'd never seen a man hit a woman before, certainly never seen two articulate adults reduced to violence. At least when you two started fighting you did it with words. As far as I know. I remembered the way Yianni took her wrist so she couldn't shield her face and I shivered. Let us not come to that. Deliver us from our instincts.

I looked round. Ruth was crawling across the floor as if she were doing a yoga exercise. She looked composed as a statue, even her hair unmoved by the wind whistling monotonously through the stones. She seemed too calm.

'You OK?' I asked.

She stopped and looked round. 'What?'

'Are you OK?'

She turned back to the earth under her fingers. 'Of course. Why not?'

I could think of at least five reasons without making any effort at all, or indeed getting onto any of her particular difficulties.

'I was thinking it won't be easy for you to go home.'

'It's not clear it's going to be possible for any of us. I've probably got least to lose.'

'Your family. Friends. Your thesis.'

Your life itself. You'd have thought she'd know more than any of us about the preciousness of life.

'I'm not worrying. There's nothing we can do now, anyway.'

I ran my hand across the floor. It was cold and gritty, and the feel of it on bare skin reminded me of the playhouse outside the church and Mom telling me not to rub my hands in the dirt when we were about to go over to the Lockies' for lunch.

'I guess that's what's worrying Yianni,' I said. 'He can't do anything about it.'

'Well, some things are like that, aren't they. We've all got to live with it.'

She came to the wall and turned back. Like a machine, a mower or a vacuum cleaner.

'Some things aren't,' I said. 'We do have some power for change.'

She crawled past me. 'Oh, shut up, Jim. Keep it to yourself, OK?'

I clenched my hands. Grief is not an unlimited license to be rude.

By one thirty, we were making good progress. We'd been through the hall as if it were a crime scene (which I suppose it almost certainly was, one way or another, at least by modern standards), and were as sure as you can be that we had left nothing we meant to take from the grave. The wind had dropped. Heavy grey cloud smothered the hill a few paces above the grave and in the silence I could hear my heart beat and my stomach growl as if I had a blanket over my head. The greying grass and Ruth's smooth hair were beaded with fog.

'Ruth? Are we done here?'

She was kneeling in the corner where we'd found the first man, holding a sieve full of earth in one hand and gazing at the dark soil she held cupped in the other.

'What? Oh, yeah. I'd say so. Did you find anything?'

'No,' I said. I'd already told her that. 'Got anything there?'

'No.' She trickled the soil from her hand into the sieve. 'Nothing at all.'

'Well, shall we have lunch then? I'm hungry. Yianni said one thirty. It's past that.'

She picked up another handful from the sieve.

'Sure. Whatever. Go ahead. I guess we can fill in after lunch.'

The soil ran through her fingers again. She bit her lip and looked away.

'Ruth? You OK?'

She blinked. 'Go get your lunch. I'll be down in a minute.'

When I got back down, Nina had emerged. She was sitting in the entrance of the stores tent, mashing something in a bowl. The smell of canned tuna carries even through an Arctic fog.

'Hi there,' I said. 'Feeling better?'

She glanced up. 'I'm OK. Yianni wants tuna again. I think we'd do better to save the protein. No harm in taking it home if I'm wrong.'

'Did you sort things out with Yianni?'

'He apologised. The next box is the last of the crackers.'

I picked up the open box, still half-full. 'Looks like Yianni calculated just about right, then.'

She looked at me, not smiling, and then turned back to the tuna. We finished the mayonnaise last week and mashing it without makes it even more like cat food.

'It doesn't need to be puréed,' I said.

She put the bowl down. 'You do it, then.'

I picked it up, and saw that she'd added more of the thyme she finds in the turf.

'When do you think the thyme dies? Or does it go on under the snow? They must have been able to use it right through the winter.'

Her shoulders relaxed a little.

'They probably used it in sheep's and goat's cheese. It goes well with game. If anyone could get one of those geese . . . And the Greenlanders would have had cloudberries, wouldn't they? A bit like cranberry sauce. I suppose barley for the carbohydrate?'

'I guess,' I said. 'We're a bit far north, up here.'

'You can cook it like risotto. Barley. Or I suppose make some kind of bannock. If you could grind it. I don't think it would rise. Roast goose with barley bread, it's probably better than stewing it. And they must have had some kind of veg. Angelica. Seaweed. Wild garlic? I haven't seen any. They had gardens, didn't they?'

'Yes. I think there's evidence for onions. And other roots.'

She gazed into the fog and shivered.

'Maybe it was all right then. Through the winter. I wish this fog would lift.'

'It will,' I said. 'Call the others?'

I couldn't see any point in separate plates, which would have to be washed by someone with bare hands in cold water. At first all these picnics were fun and the blandness of the food meant we were serious archaeologists, sacrificing ordinary comforts for our work. Now I just want a hot meal at a table, with chairs and lights and enough warmth that I can take my gloves off and still control cutlery. I want good food as well, sweet, salty, spicy, fresh food. Food from that Vietnamese place I took you to, Mom, when you visited last time, salads with lime and chilli, seafood in crisp little parcels. Then I want serious desserts. Mom's peach cobbler. Lemon cheesecake. I remember liking fudge sundaes but can't, at the moment, imagine wanting anything colder than myself. Hot coffee. Hot soup, chowder with cream and bacon and clams and snipped chives floating on the top. I should stop this. I'm like a condemned man obsessing over his last meal when eternity begins at dawn.

I watched the others' faces when they saw the meal. Yianni's glance flickered past the plates, up to the fog and out towards the veiled water. Without wind, the waves had fallen silent and nothing moved.

'We'll have to be careful not to miss anything,' he said. 'I was relying on being able to scan the site.'

'Yes,' I agreed.

He sat down and sighed. The others came from the chapel. Catriona seemed to have lost her hat, and her hair was slicked over her forehead and dripping down her back.

'Yianni?' she asked. 'Have you checked again? The phone and the computer?'

He shook his head.

'Well then, can we? Please? Maybe you'll get through.'

'After lunch, OK?'

She stood shifting from foot to foot, arms wrapped around herself. 'Can I have a go on the computer, then? Now? I'll only try to connect. I won't change anything. I'm not hungry.'

'I don't want to use up the battery,' said Yianni.

'Why not?' said Ben. 'If we're leaving, it doesn't matter. You can plug it in on the plane home if you want. Though, personally, I'd rather watch the movie.'

I looked up, trying to imagine that somewhere above the weight of cloud there were planes with people sitting in upholstered seats reading shiny magazines and eating salty snacks, fretting because the seats don't recline far enough and the baby up front can't get to sleep. (Worrying it might fall out of the sky, or worse. Scanning the other passengers between movies and sleep. I wish we'd flown when I was a boy, when the excitement of wheeling above the clouds made the check-in queue where we left Aunt May seem more glamorous than other queues, and even the cars in the airport parking lot wore a special kind of abandonment.) People who are too warm, waiting to be fed chicken and vegetables and fruit and bread and butter. I thought that if I get out of here, I'll always remember what it feels like to be cold and hungry and I'll never complain again, but I guess it's more likely that if I get out of here I'll tell other people who are pitying themselves what it feels like to be cold and hungry and to envy people strapped into airline seats for ten hours at a time.

'It won't work any better for you than for Catriona,' Ruth

was saying. 'Give her ten minutes if you're worried about the battery.'

Catriona hugged herself.

'Oh, OK,' said Yianni. 'Ten minutes.'

Defeat. He closed his eyes and massaged his forehead with his fingertips. Catriona unfolded herself, took the laptop from his tent and sat on the wet ground at the entrance.

'Keep it dry,' said Yianni.

She opened it and we all waited while it whirred and sang. At last it played a chord, the chime muffled by the eddying fog.

'Look,' she said. 'It seems to be working.'

She opened Internet Explorer and clicked connect. The machine cogitated audibly. We came closer. She typed in Edinburgh's web address. The timer appeared, and stayed. And stayed.

'Try refreshing,' said Nina.

'Not yet,' said Catriona. Her fingers drummed beside the keypad. The computer said it was connecting to edinburgh.ac.uk. 'See?'

The window went blank, and then the 'site temporarily unavailable' notice appeared. Her shoulders rose and she typed in 'guardian.co.uk'. We waited again, as if the computer might let us wake up and find that it had all been a dream. Virtual sand trickled.

'Only a few minutes more,' said Yianni.

She clicked refresh. I pushed my finger through a hole in my glove. How much time do we spend waiting on machines?

'We should be having lunch,' murmured Yianni, but he didn't move from the screen.

Temporarily unavailable.

'It's not going to work,' said Ben.

'One more.' Catriona swallowed and closed her eyes for a moment.

She typed 'hebrideanestateagents.com'.

'What?' I asked. 'Real estate?'

'Imaginary estate,' said Nina. She rested her hand, blue and blotched in fingerless gloves, on my arm. 'She's thinking about buying a croft.' She shivered and put her head to my shoulder. I wondered whether to put my arm around her, and as I lifted my hand something hit the tent behind her.

'Fuck,' said Catriona. 'Fucking hell, what the fuck was that?'

(Sorry, Mom. British women swear.)

'Jesus Christ,' said Ben.

Nina raised her head, looking not at the computer but away up the hill, from where the missile had come.

'Look.' She pointed.

Maybe something moved through the fog. Maybe the fog moved through the rocks. Fear hammered on my sternum. I told myself the psalm, more of a ritual than a prayer. The Lord is my light and my salvation; whom shall I fear? The Lord is the strength of my life; of whom shall I be afraid?

'Now do you believe me?'

'There is someone else here,' said Ben. 'Someone who doesn't like us. Jesus.'

'Yes.' Nina was still gazing up through the fog. 'They weren't ready to die. And you've been robbing their graves. Of course they don't like you.'

'Shut up.' Ruth stood. 'Dead people don't throw stones. It fell. Stones do fall, remember? On hills? That's how they get into valleys.'

Yianni bent and picked up the stone.

'It's from up the hill,' said Ben. 'It's the right kind of stone.'

Catriona buried her face in her hands. 'I want to go home.'

She sat there, shoulders heaving. I looked behind me, quickly, but there was only fog.

'Tomorrow,' I said.

'Maybe a bird dropped it.' Ben took the stone from Yianni. It was the size of a baseball.

'Some bird,' said Nina.

Yianni went round and stroked the tent as if it were an injured animal.

'Stones fall. You've all got so nervous you can't think straight.' Ruth knelt on the turf with her back perfectly straight.

Yianni picked up the computer. 'I'm going to put this away now. It's got all our data.'

'You've been backing up,' said Ruth.

'Always. But I'm not taking risks.'

He took the machine on his lap, folded it into his coat and crawled into his tent.

Ben stared in the direction Nina had pointed. 'I don't like this.'

'We're going tomorrow,' I said. I think that was when I stopped believing it.

*

Yianni wanted us to use the remaining hours of daylight to search the site and the camp.

'Pick up everything, OK? Any shreds of tissue paper, even tiny scraps of food wrappers. Pen lids, labels. I'm serious. Come spring, the rangers are going to be out here checking the site and it has to look as if no one's spent a night here since the Greenlanders left.'

'Yianni, we know. We're archaeologists, remember? We know how hard it is not to leave traces.' Ruth was swinging her arms. The fog was thicker and the temperature was falling. I couldn't see across the river and I thought about how bush planes don't fly in fog.

'You removed six bodies and all their surviving possessions,' said Nina. 'I don't think you can pretend nothing's changed. You've only left the buildings.'

'Whatever.' Yianni looked away, up towards the screes where the big stones were. 'That's what we're doing, OK? Think of archaeology as being like mining, Nina. Or surgery. Whatever you take out, you tidy up the surface so no one can tell. Please, Nina. Just help for the last few hours.'

'Sounds more like burglary to me.'

'Just do it.'

Catriona stood close at my elbow. She kept glancing round. 'Yianni? I'm really sorry but I'm too scared to go around on my own in this fog. When it seems there's someone hiding in it.' She shivered.

Yianni looked at her for a moment, and then he too scanned the fog.

'It's better to work in pairs anyway,' he said. 'Less likely to miss something.'

'I'll go with you,' said Nina. 'We'll stay away from the grave.'

This didn't seem at all reassuring, but Catriona nodded and joined her.

'Nina, please,' said Yianni. 'You can shout if anything worries you, can't you?'

Stones do fall, of course, but usually something pushes them. There was no wind. I could think of a lot to shout about.

The fog was still there when the dark came back. It was as if light was being blotted from the air around us, and when Ben and I came back to the tents I tripped over a cup which did not spill. Ice. It seemed worse, somehow, more alarming, in the domestic context of cooking and eating and drinking than down by the river. You expect the great outdoors to freeze, but not your kitchen. I remembered Grandma talking about having to melt water to wash her face in the mornings and waking that time in the new house to find snow on her quilt. Only two generations ago, and I guess in another few decades we'll be out of fuel for the central heating anyway. Assuming global warming doesn't see to it first. Grandma would have done just fine out here, and for all I've worked on the Greenlanders I bet she'd know more about what it was really like, cosseting the animals through the winter and going hungry in the spring, heating water on an open fire when you wanted to wash your clothes and going outside for the bathroom.

'Jim?' Ben was blowing into his mittened hands. 'This fog. The plane couldn't land anyway, could it?'

'I don't think so. Can't see how it would. It's not as if we've got radar or landing lights on that field. It might lift. Well, it will lift. Eventually.'

'Yeah. That's what I thought.'

Nina cooked that night. Last night. Catriona, who returned from her afternoon with Nina with a graze on her forehead, found a tube of tomato purée, mis-categorised with the antiseptic creams in the first aid kit. Then she rummaged in her tent and came back with a small tin of anchovies.

'I've been saving them.' She handed the tin to Nina. 'The nutritional value's minimal, we won't – I mean, if the plane can't land, we needn't regret eating them. My mum gave them to me when I was packing. She used to take them when we went camping when I was little.'

Nina, face hollow in the lamplight, looked like someone in a Christmas crib painting beholding the baby Jesus.

'I can save half. Or more. They'll keep, in this cold. Thanks, Cat, they'll make all the difference. I picked some thyme.'

I think it will be years before I want thyme again, and maybe it will always remind me of prickly grass through my jeans and the fleeting warmth of a metal plate in my hand. Hands and face moving through the small circle of yellow light, Nina squashed garlic in a dish with a fork, peeled back the lid on Catriona's anchovies and forked them dripping onto the garlic. My mouth watered. Steam from the noodles merged with the fog, and already the tin was beaded with damp. I know we are not very hungry, not by historical standards. People work for months and years and raise kids and walk miles with hunger worse than this. I don't know if you get used to thinking about food all the time or if it stops after a while or if other people are better at lifting their thoughts. All those fasting saints, without noodles and anchovies or even thyme.

'It's nearly ready,' said Nina. 'Yianni?'

Yianni was in his tent with the phone. Silently.

'Yeah.' He sat in the entrance. 'Sorry. There's still nothing. No dial tone.'

My boots are worn. I flexed my foot and watched the crack across the toes open and close.

'I don't think we thought there would be.' Catriona's voice was high, as if from far away.

'It might be the phone. I never tested it.'

'We guessed.'

The stove fell silent. Nina took the pan off and, using the lid to hold back the noodles, poured the water unsteadily into a mug. She shook the noodles in the pan. It hissed as she put it down on the grass, and a faint smell of hay rose. Her hands in the lamplight were misshapen by chilblains. She scraped the crushed garlic and anchovies over the noodles, squeezed a worm of tomato purée over them, and stripped thyme stalks of their leaves. It smelt good. It smelt like real food. She stirred it, lifting the noodles through the sauce.

'There. It's only flavoured starch, really. Olives would help a lot.'

'It looks really good,' said Catriona. 'Thank you, Nina.'

It was good. It would have been a great entrée, followed by steak and potatoes, a salad, apple pie. Nina had only taken a couple of mouthfuls and mine was nearly gone. I tried to slow down. Fog eddied between the tents and Catriona watched it.

'It's just fog,' I said.

'I know.'

I could hear Ben chewing. Ruth's fork clattered on her plate. I remembered again my dream, with the light from the window falling on snow outside and Hannah waving her spoon. Ordinary life, no portents. We were talking, all of us, and I think when I woke up I could still remember what we were saying. It's fading now. I looked around at the hands and faces glowing out of the dark. I will not miss these people. Would not. Ben caught my eye.

'OK.' He put down his fork. 'So, what do you think happened to the Norse Greenlanders? The whole colony. Why are they lost? I don't mean what do you say at conferences or to students or whatever, I mean really, what do you see?'

'I think raiders came,' said Nina. 'Like in Ireland. There's a village I stayed in once, and Barbary pirates came in the sixteenth century and kidnapped everyone. It's the same landscape. We're only further up the same coast, geologically. They were here anyway with the cod fisheries. They took the livestock and the young people for slaves and the old ones watched the ships disappearing over the horizon and knew the future was gone.'

'There are problems with that argument.' Ruth twirled noodles around her fork and inserted them as if she were wearing lipstick.

'I know,' said Ben. 'There are problems with all the arguments. I just want to know how you imagine it.'

'You mean you want a story?' asked Nina. 'I thought you didn't like fiction.'

'Not all stories are fiction. Go on, Ruth.'

'Climate change,' she said. She began to wind up another mouthful. 'Colder and hungrier over a generation or two. Higher mortality, harder lives. If they'd been able to adapt they might have been able to stay but they couldn't, any more than we can. Going on with saunas and trying to grow grain.'

'Like the Americans with those enormous cars?' Nina put her plate down, half-full.

'And the Brits. I imagine the harvest getting later and less each year, and people talking about what their grandparents had been able to grow and trying to figure out what was different. Thinking the winters seemed worse and the summers shorter and not knowing if that was just how memory works.'

'They'd have known,' said Catriona. 'Wouldn't they?'

'Not everyone. The change was slow. Slower than now. And they wouldn't have known if or when it was going to change back. People are pretty conservative, you know. They don't change until they don't have a choice.'

'So what about the end?' Ben had finished eating. 'What about when it came to the crunch?'

Ruth chased the last noodles with her spoon. 'They went out with a whimper. Back to Iceland. Maybe some outbreaks of plague. A few pirate raids. Malnutrition. There must have been some intermarriage with the Inuit, it's not credible that there wasn't.'

'It hasn't changed what you think,' said Nina. 'This dig hasn't changed anything.'

Ben put his plate down. 'It was never going to, Nina. Were you thinking we'd find the answer?'

She shrugged. 'What do I know, I'm not an archaeologist.'

'Oh, Nina,' said Yianni. 'You wanted an ending. It's just evidence. More evidence. One way or another.'

I remembered the stories I used to tell the girls in the back of the car. The family of beavers who got swept down the river one day and out to sea and, after adventures that got us across two states, made a new home on an island in Puget Sound. Do you remember how after Hopper got run down, the beaver stories kept Hannah calm enough to go to sleep?

'I think they sailed away,' I said. 'I think they packed what they could take on a ship. Not the big stuff, like looms. Just tools so they could make the other things again. They were sad to leave, when they'd lived in these houses so many generations, and especially to leave their churches. And their dead. But they'd known about Vinland for a century or two, and when the time came they knew where to go. So they slaughtered most of the livestock, as if it was fall, and dried the meat. They took the strongest animals alive, because they didn't know if they'd find sheep and hens at the other side of the sea. They shut the doors and walked away from their houses, and they sailed away, not much more than one family to a boat. They got really sick on the voyage, and they hit some really bad weather –'

The lamp guttered.

'Wind,' said Yianni. 'The fog's lifting. Go on.'

'But nobody drowned, and after a couple of weeks they saw land on the horizon. It was warmer than they were used to, warmer than any of them could remember, and they unwrapped the blankets from the children. As they got closer in, they heard birds again, and soon they could see tall trees. They'd never seen trees so big and green before, and they were so happy to think what they'd be able to build with timber like that. Nearer yet, they could see big breakers on sandy beaches. They had to anchor overnight, and then work their way along the coast looking for somewhere to land, but at last they found a bar and behind it a natural harbour, where the water lay in the sun like – like kitchen foil.'

'They wouldn't have known about kitchen foil.' Nina had clasped her hands round her knees.

'And it's too shiny. Never mind. The water lay quietly.'

'That's better.'

'Thank you. They sailed over the bar and into the harbour, worrying a bit about the tides and how to get out, but tired and so eager to set foot on the green land –'

'They'd have had to row,' said Catriona. 'If the water was that flat. No wind.'

'And as they crossed the bar the wind dropped, but they had the oars ready and, with the little boy taking soundings, they rowed slowly to the bluff and brought the boat alongside, where it bobbed comfortably with five feet of water under the keel.'

'You said it was flat,' muttered Nina.

'And then they climbed ashore. It felt strange to have their feet on the ground after weeks in the boat, and they walked unsteadily. The men told the women to keep the children close to the ship while they, with their battle-axes ready, scouted for wild animals or hostile inhabitants. They knew the stories about encounters between earlier Vinland voyagers and the Skraelings. It was hard for them to check because they were used to the open landscapes of the fjords and the forest unnerved them, but they saw no movements but the flight of birds and heard no voices but their own. The sounds of the forest were strange to them, people who had never heard the wind in the trees or seen sunlight filtering through leaves. They returned to the ships to find the women spreading fishing nets and the children playing under the trees, already making a playhouse of fallen branches –'

'What?' said Ben.

My scalp was prickling. Nina gazed intently up the hill.

'Someone's listening,' she said. 'Not far away.'

Cold crept up my spine. Catriona whimpered, a hurt puppy, and moved suddenly back into her tent.

'Stop it,' said Ruth. 'Go on, Jim.'

I leaned forward and looked round, behind my tent and up the hill. The darkness was absolute. Even here, there's usually something, starlight or moonlight or some kind of sense of where things are. Humans, after all, have not had electric light long enough to lose whatever instinct we had for avoiding predators in the dark. The lantern flickered again.

'Maybe we should put it out,' I said. 'Save fuel.'

'They still know where we are,' said Nina. 'I've got lots of torch batteries. Long life. We won't be left darkling.'

'Go on,' Ruth said again. 'Finish your story.'

I remembered the beaver family, the way the serial ran all

summer until I couldn't stand it anymore and sent them on a one-way trip to the moon. Let this not be a serial. Let us be home soon, with half a continent and an ocean between me and the blonde seer of ghosts.

'I don't know,' I said. 'I should think they built a house and lived happily ever after.'

'What, like Roanoke?' said Nina. 'Is anyone still looking for the lost Viking Americans? Didn't someone claim to have found them in Appalachia?'

Ben, who had been leaning back on his elbows, sat up. 'People find all kinds of weird shit in Appalachia. Nice story, Jim. Thanks. I guess this was what people did before TV.'

Nina grinned. It made the sinews stand out in her neck.

'We can always read aloud,' she said. 'All winter, if needs be. Shame Ruth hasn't brought her needlework.'

'I don't have needlework,' said Ruth.

Something whistled, up on the hill. Two notes, rising. Some kind of call. My throat burned and my mouth filled with the taste of sour anchovies. Nina was staring at my face.

'You heard that.'

'A bird,' I said.

'Heard what?' asked Yianni. I looked round. Catriona in her tent was just a ball of darkness in the uneven glow from the lamp. Ruth was gazing into the lantern and twisting her hair round her finger. Ben looked away.

'Someone whistling,' said Nina.

Ben looked up. 'OK. I heard it too. It wasn't a bird.'

Nina's face changed, cleared. 'Really? You heard it? You really did?'

Ben's shoulders hunched and he pushed his hands inside the opposite sleeves.

'I did too,' I said. 'I don't like it.'

'Thank God,' said Nina. 'So you don't think I'm mad?'

'I don't know. It seems like there's – something – in the valley with us. I don't want to meddle. Maybe Ruth's right, we're just scared and a long way from home.'

'They're doing the meddling,' said Nina. 'Oh, good. So you all believe me now?'

'I don't,' said Ruth. 'And I don't believe either of you. I didn't hear anything. It's like seances – you get a group hysteria where people convince themselves something's happening. I don't want to hear any more about it. Have you thought what these ghost stories are like if you've actually lost someone? I'm going to bed.'

She turned round, crawled into her tent and closed the zipper.

'Hard to storm off into a tent,' said Nina.

Catriona poked her head out of hers. 'You should know.'

They began to giggle. It sounded as strange as the computer chime in the darkness and the cold.

'OK,' said Yianni. 'Come on. Let's all get to bed. I'm setting the alarm for six. So we'll be ready if they come at first light.'

The giggle died.

'The thing is,' said Catriona. 'I'm scared. I'm so scared of what might happen in the night.'

'Come in with me,' said Nina. 'I mean, if you want to. At least there'd be two of us.'

'Do you mind?'

Nina looked up. 'No. I've hated being in the dark on my own. Only my books'll have to sleep in your tent.'

Yianni sighed and stood up. 'Come on, then. I'll help you move the books.'

'What about you?' I asked Ben. 'You OK?'

He looked as if I was propositioning him.

'I'll cope. Thanks.'

I guess he'd rather face the Texas chainsaw massacre than admit either he's scared or that two straight guys could comfort each other if they were to be scared.

'Goodnight, then. Sleep well.'

I lay in my bag again, shaking with cold, my neck muscles twanging when I tried to move. I don't know about the whistler and the thrown stone. They were there. I wish they weren't but they were. They are so slight, a stone falling in a stony place, air through something narrow. Nothing you'd notice anywhere busier. But they don't feel good. I lay there in the cold and dark and tried to pray.

The alarm shrilled, and I tried to reach for it and couldn't. Sleeping bag. Tent. I should know after all these weeks. Not my lumpy bed and Mom's quilt, which are cold and dusty at Allen Street. Not my alarm but Yianni's. The darkness was so dense I hoped for a moment it was still the middle of the night and we could go back to sleep, but a muffled glow like Holly's nightlight came from the left and Yianni coughed.

'Morning, everyone. Time to get up! Everyone sleep OK?'

I could hear people stirring.

'It doesn't look like morning,' said Catriona.

'Just gone six,' said Yianni. 'Sunrise in a couple of hours. They might come any time after that.'

Someone yawned.

'We can't take the tents down in the dark,' argued Nina.

'There'll be some light. If you can eat dinner in the dark, you can eat breakfast. Come on.'

I sat up. I could almost hear the muscles creaking. I pushed down the hood of my bag and cold bit into my face and ears. I pulled it up and lay down again. Nina's torch came on and she and Catriona started bickering about which of them was going to dress first. It was like hearing the girls on a school morning; if you dress first I'll have time to wash my hair, if I have the bathroom now you can get the shirts from the laundry. I hope they're still doing that. I hope you're all still there, Mom, still pottering about until seven-thirty and then rushing around till the school bus comes, as if the passage of minutes in the morning were a daily surprise.

'Jim? You getting up?'

'Yeah,' I said. 'In a minute.'

Just like I used to say to you every morning. I sometimes wonder why you bothered, Mom. Did you never think about leaving me be, just letting me be late and take the consequences? I tucked my hands into the warmth of my armpits and felt the chill through my thermal T-shirt and polo-neck and two sweaters.

'Minute's up, Jim.'

I sat up again. I thought about being on the bush plane and watching the valley drop away and vanish, the fall of scree and the river and the beach, the field where we've camped and the fallen stones of the chapel and homestead passing in seconds and becoming indistinguishable from all the other fading valleys and freezing inlets out here. I started dressing, thinking about landing at Nuuk. About using the men's room without howling cold and getting a hot coffee from a machine and maybe even an English-language paper to read while waiting on a real chair for a bigger plane, one with reading lights and flight attendants and passengers I don't know, to take me to Copenhagen. Where there are cafés at which a person might choose to spend money, even, after so long away, enough money for a Danish coffee, and the *New York Times*, and you can see trees and sculpture and parking lots and crowds of people. Maybe. If crowds are something people are still prepared to risk. If airports are still open. If there are still enough people to make crowds.

I took a deep breath and opened the tent and crawled out. The next breath felt like a bruise to the chest, like the first morning after a blizzard when it's forty below and you step out of the kitchen into the yard and your body doesn't quite believe it. Yianni had the lantern going again, and he was kneeling on the ground with the stove between his knees. Catriona's and Nina's heads peered out of the pink tent as if it were one of those Chinese New Year dragon costumes.

'Cold.' I tried to stop my teeth chattering.

'Not so bad for the time of year,' said Nina. She and Catriona giggled again. It's like when the girls are in a silly mood and it's getting tiresome. You'd think she might find ghosts we're all beginning to believe in more sobering than her own delusions.

Ruth's zipper went. She breathed out and a small cloud hung in front of her face. She pulled her scarf up over her head like an Italian grandma going to Mass.

'Coffee?' she asked.

'Just fruit.' The stove flared and Yianni knelt back.

The pink Hydra disentangled itself and Catriona crawled out.

'Yianni? How much fruit is left?'

'More than enough.'

She blew on her hands. 'But how much? Because it's our only vit. C, isn't it?'

'There's enough.' He turned to pour water, put the pan on the stove. 'We've got vitamin tablets.'

You used to tell me they were growing pills, do you remember? When I was worrying because I wasn't the tallest in the class anymore? You said they were pills to make sure children grew just as tall as God wanted them to be, and they had to be taken with milk, which left me with the idea that somewhere in the Bible God showered his chosen people with Vitalife, maybe along with the milk and honey. Milk, incidentally, inhibits iron absorption. Orange juice would have been a better bet. Right now I'd give a lot – not that I've got a lot, of anything – for either.

By the end of breakfast we could see a faint outline of the mountains against the sky. I have come to know each rock. I look along that skyline and know what makes each hump and decline. There was a new point on the hill-top.

'Looks as if someone built a cairn in the night,' said Nina. 'See? And this time you know it wasn't me, unless you think I could have climbed over Cat and got up there in the dark.'

'You said you had lots of torch batteries,' said Ruth. Her scarf shadowed her face.

'Still do. But you can't think I'd climb up there in the middle of the night to build a pile of stones?'

Ruth shrugged in the half-light.

'How do I know what you'd do?'

'How do you know it's a cairn?' asked Ben.

'We robbed their graves. They build memorials. That one's where they can watch us every hour of every day until we leave.'

'Which could be in two hours. Come on, Nina. Enough.' Yianni stood up and stretched. His bones cracked and Ruth winced. 'So. Washing up first. Pack the stores, pack your belongings. Tents down. Everything over the river ready to load when the plane lands. Final search of the camp.'

'Are you going to try the phone again?' Catriona was scanning the mountain top. 'Is there – no. Never mind. Oh God. Look.'

We looked.

'What?'

'I thought I saw something. Moving up there. By that – cairn. Someone watching us.'

'Well.' Ruth stood up. 'If there is, maybe they'll be able to tell us what's going on. But I can't see anything.'

'Me neither.' But Ben glanced around, behind the ring of faces.

'I told you,' said Nina. 'It's just a matter of whether they move in before we get away. At least it'll be fully light soon. We'll be able to see.' She shivered.

'It's whatever it is seeing us I don't like,' said Catriona.

'Oh, for God's sake. If you give me a torch I'll get some water and wash up.' Ruth banged the plates together and took the bucket down to the river.

The hardest bit was ferrying everything across the river. We're all used to the stepping stones by now, but it's different when you're carrying things. The tents and rucksacks were OK, slung on our backs, but the last bits of food and all the finds were packed in boxes. We formed a row across the stones – the closest collaboration, it occurred to me, since we arrived – and then we were left with the five long boxes.

'If you drop any of these I will kill you and bury you at sea and say there was a freak polar bear encounter,' said Yianni.

We were still strung out across the stones. He stood on the far bank, next to the pile of boxes, which looked, in the end, smaller than the boulders grazing the field.

'Don't.' Nina, straddling the shore and the first stone, shifted her weight. 'I don't want to be scared of you as well.'

Their eyes met for a moment and I looked away. There are, however, records of the Greenlanders encountering polar bears in the winter. We are far enough North.

'Yeah. Let's come off the stones. We need to carry these guys together.'

'Pall bearers,' said Nina. 'Jesus, I hope that plane comes.'

Ruth stepped back onto the shore beside Yianni.

'I'll take one end,' said Yianni. 'Jim? You're more my height.'

The Greenlanders must have had the same problem. The lucky ones died in the summer when burial was easier in the soft ground and falling into the river less of a problem for the bearers. It was Yianni, not me, who fell in, and although he swore till even the watchers on the hill must have turned away he did not drop the box. It meant we had something to do when everything was stacked on the far side, trying to warm him, digging out a change of clothes and making a drama of getting the girls to go behind a rock while he changed. Two rocks. Ruth and Nina can't even share a boulder.

By 7.53 we were all sitting on our backpacks on the flat ground above the river, close enough to the chapel that I felt uncomfortably overlooked. I gazed back towards the campsite, where rectangles of etiolated grass were all that was left of our settlement, and understood why farming peoples are so interested in the anthropology of nomads. It must take a different kind of confidence to leave no signature on the land. We will have left something, something a good forensics team would find and a good archaeologist be able to interpret, but actually the most obvious sign of our presence will be what we have taken away.

'That should be visible from the air, anyway,' said Nina. 'Until the grass grows back.' She was following my gaze.

'They know where we are,' said Yianni.

He looked at his watch, and then up at the northern sky. From whence cometh our temporal salvation, if at all. The fog was gone and the sky was white and blank as a dead screen.

'Did you try the phone again?' asked Catriona.

He didn't meet her eyes as he nodded.

'Oh.' She fiddled with the zippers on her backpack.

'So what's the plan?' asked Nina. 'We wait all day and then when it gets dark, what? Put the tents back up?'

I concentrated on the plane out of Nuuk, being pushed back into a padded seat as the nose lifts and the aisle tilts, the pretty girl who offers me a drink when the seatbelt sign chimes off, the free magazine with pictures of shiny people doing shiny things in Copenhagen.

'The *plan* is that the plane comes and we get on it and get the hell out of here, isn't it?' Ben was already drumming his toes on the ground.

'OK.' Nina was enjoying this like it was Debating Club. She is a member of something called the Oxford Union, which is not a union. 'What's plan B?'

I heard your voice, Mom. 'Let's not borrow trouble.' Sufficient unto the day. I don't think I'd ever said it before.

Nina stretched. 'There's a difference between borrowing trouble and making a contingency plan. In a few weeks, we might really regret wasting a day of light waiting for something we kind of knew wasn't going to happen.'

Yianni stood up and I tensed, ready to leap in this time. He turned and walked a few paces towards the river and then back.

'OK. I agree there's no point us all sitting here for – as long as it takes. Go. Do whatever you like. Just don't put more than twenty minutes between yourself and this field and turn back the minute you hear the plane.'

'OK,' said Ruth. 'I'm going for a walk. To see the next valley. I'll call you if there's a troglodyte settlement.'

She began to walk fast up the river bank, leaping from one clump of reeds to the next.

'Don't go too far,' called Yianni. She didn't look back.

The rest of us sat a moment. 8.04. Around eight hours till dark, though I presume they'd aim to fly back in daylight as well. It's not that far. Half an hour, Nina says, in a plane smaller than a car, where you can see the pilot turn the windscreen wipers on. Nina came on the plane because Yianni didn't want to subject her to horses. So the latest it might come, if it is coming, would be about 3.30. Seven and a half hours.

'Come to the beach?' Nina asked Catriona. 'We can look for the perfect shell.'

'Don't take anything from the beach.' Yianni seemed to be talking to his boot, which he was rubbing across the grass as if he'd trodden in dog dirt.

Catriona shook her head. I don't think she's been to the beach since a wooden dinghy blew ashore one night a couple of weeks ago. Nina saw faceless pirates in it and Catriona got scared.

'I'll do some painting. I packed my paints on the outside in case there was a wait.'

Nina shrugged. 'OK. I'll stay here, too.'

'I'll go with you,' said Ben. 'Might as well stretch our legs. Could be a long journey home.'

'Could be,' said Nina.

They left. Yianni stared into the horizon again. 8.13. I don't know if you'd see a plane first or hear it. I guess it depends on light levels and wind direction, but the only sound here now is the wind. My guess is we'd hear almost anything happening within a good few miles. Yianni unfastened and re-threaded a strap on his backpack.

'Are the field notes all finished?' I asked.

Catriona was leafing through her artist's block. Landscapes in shades of grey flashed like cartoon images.

'Yeah,' Yianni said. 'I might look through them. Make sure it's all complete. Maybe a few more sketch maps. While we're here.'

He looked over the valley again. No wonder the Greenlanders didn't make maps. Land must have divided into what they knew as well as their children's faces and what was entirely extraneous. Little reason and scant means to travel overland outside their own farms. Yianni rifled through his laptop case. I looked up the valley. Ruth was climbing fast, a red and green insect scurrying towards the loose stones. The new cairn was still there, black against the pale sky. I wondered if she was heading that way.

Out at sea, white waves rolled down dark water. When I sat quite still, silencing the rustle of my coat and the scratching of my fingers staying warm in the pockets, I could hear the higher notes of Nina's voice and then, to my surprise, Ben's laughter. Nina appears, some of the time, to be back with us, the way she was in the beginning. Now other people are seeing her visitants. And Ben's good with her, good at not being scared. I glanced back up to the cairn. Ruth had reached the bottom of the scree and slowed down, but it looked as if she was still closing in on the cairn rather than aiming for the next valley.

8.21. I pulled my Bible out of my pocket. The one

Grandma gave me. It's getting worn. I once heard Hannah ask her if penguins went on the Ark or just swam through the flood, and they went off to her bedroom to look it up and see what the Good Book said. Goodness only knows where Grandma found a text for that one. I remembered her reading the Christmas story to us all on Christmas Eve that last year, after I read *The Night Before Christmas* to the girls and even Holly pretended to look for reindeer in the yard. If we are still here at Christmas . . . Stop it. Don't think that. I hope Hannah knows she's still young enough for *The Night Before Christmas*.

I started at the beginning of Matthew's Gospel. I kept seeing the church at home, the shiny wood of the pews and the holly and ivy and red ribbons around the altar and under the windows. The candles standing tall against the white walls and the gleaming cross up front. That time Holly realised on Christmas Eve, the first time she could really read the carols, that the Baby Jesus and Christ crucified were the same person. He went back to God, you told her. He came here to be with us and then he went home. What about his mom, she kept saying. What about the Baby Jesus's mommy watching him being nailed up? Not the Baby Jesus, you said. Grown-up Jesus. Nearly as old as Daddy. Now then, where shall we hang your stocking? She didn't mention it again. Do you remember? You went off and talked to the pastor about children and the crucifixion and he gave you a bunch of books that lay around the kitchen right through Lent, mostly under the pile of letters and Hannah's drawings and forgotten shopping lists, which I guess is still on the kitchen table. Come Easter we were all primed and all she was interested in was the egg hunt. And now it's chasing boys.

It's almost hard to read those parts of the gospels now, I know them so well. My mind skipped ahead, from the innkeeper to the angels, the shepherds to the sages. Herod, and the massacre of the innocents.

At 8.51 Nina and Ben came back. I saw her look quickly into the northern sky and then up the valley. Ruth was now working her way along the shoulder of the hill, towards the pass.

'Nothing's happened,' I said.

'My painting's happened.' Catriona shook back her hair and looked up. She'd painted me and Yianni and the boulders, all sitting hunched on a wash of grey-green punctuated by clumps of reed. The sky was blank.

Nina smiled. 'That's lovely. It's like those Norse folk tales where you can't tell what's a tree and what's a troll.'

'It's not finished.' She dipped her brush in the film canister of water she carries and began to add the curve of the river to the background.

Ben sat down near me. 'What've you been doing?'

'Not much.' I put the Bible away. 'How was the beach?'

He shrugged.

By 9.30 Ruth had disappeared into the next valley.

Nina looked up from *Romola*, another George Eliot which I remember Rachel saying was unreadable when she took a Victorian Women Writers course. 'She must be more than twenty minutes away.'

Yianni was still reading through his notes, though I'd been watching him for a while and he hadn't turned any pages. He didn't look up.

'There's not much I can do, is there? I'm sure she'll come back if – when – the plane comes.'

'We wouldn't know if anything happened to her,' Nina persisted. 'You told her not to go out of sight.'

Yianni looked up. 'You want me to go after her?'

'I'm just saying.'

Catriona put her sketchbook down and weighted the top page with her paintbox. 'Pass me a couple of those little stones, Jim? By your foot?'

She tore out the sheet and laid it carefully on the flat of her backpack, a pebble at each corner. Yianni turned a page and noted something in the margin. Nina had gone back to medieval Florence and Ben was leaning against his pack with his eyes closed. His breathing was too fast for sleep.

'I might go up to the hall,' I said. 'Have a final look round.'

No one replied.

I was half hoping to find a bootprint or a candle end, some sign that the other person in the valley had a physical mass and modern accessories, but the hall was as quiet and empty as the day we arrived. I stood in the doorway, where we found a scattering of small, sharp objects dropped by people taking fiddly work to do in the light. Catriona said that on some North Atlantic islands there seems to have been a late medieval tradition of burying a baby or small child under the threshold and Yianni said he knew of no Greenlandic examples. Doesn't mean there aren't any, but we didn't lift the huge flagstone at the door. Unless the shoreline has changed a lot, they'd have been able to see a long way out to sea from here. Women watching for sailors home from the sea, or just the men bringing home fish, which would be pegged out on wooden racks to dry against the winter. We have nothing, you know. We've done the opposite of what the Greenlanders did. We've spent the summer eating all our provisions, and the only things we've harvested are dry bones and pots dropped and smashed five hundred years ago.

I stepped outside to look at the sky. Nothing. I scanned the sea, as if someone might come that way. As someone did, long ago, square sails rising over the horizon, telling the people at home to kill the fatted lamb and open the best mead. That, or run for the hills. I looked along the shoreline, concentrating on the point where the land curved away into the next inlet, the place where the shepherds emerged into our valley. People can travel along there. Surely if it can be done on horseback, it could be done on foot?

Something caught the back of my coat. I turned, glimpsing, I guess, my own shadow, and in the next room a stone rolled. I didn't want to be there anymore, and I set off down the hill, trying not to look behind me, thinking about coming home for the weekend, sitting round the table and trying to describe this place to you.

Catriona was still painting. Nina was still reading. Yianni was still staring at his notebook. Ben had got his plane ticket out and was colouring in all the o's on the back of the folder.

'Checking the date?' I asked him.

He looked up and reddened. 'Just checking. I hadn't seen it in a while.'

'Passport?'

'Yeah. Here.'

Nina looked up. 'Gosh, that looks very new and shiny. Mine's in rags.'

'Yeah.' He pushed it back into the wallet. 'The other one went through the wash.'

'Oh.'

She went back to her book. Ben went back to his colouring. Catriona had arranged some pebbles in a nest of reeds and was painting each reed individually onto her pad with a brush the size of a matchstick. She stippled dark brown stripes on the dead yellow-green. Ruth was invisible still and I began to wish I'd gone with her. 10.27. I thought about heading off on a walk of my own and remembered the hand plucking my coat. I thought about checking my own passport and tickets, as if I hadn't put them into the top of my rucksack first thing that morning, sealed into a Ziploc bag in case of rain. I thought about waterproofing my boots. I thought about trying to read some more of Matthew. I thought about trying to shave, so I wouldn't look like a tramp on the plane, even if I can't help smelling like one.

'Oh, for goodness' sake,' said Nina. 'Have a book to read.'

'What?'

'I can't concentrate with you looming and fidgeting like that. Have a book.'

'I've got a book.'

'Have one you feel like reading. I can offer you *Persuasion*, but you won't like it. *Villette*? It's a good distraction, set in Brussels. Lots of interiors and cooking.'

'Is it still about someone getting married?'

She looked at me as if I were a particularly foolish student.

'No men attended weddings in the making of this book. She doesn't get married. The anti-heroine does but it's no big deal.'

I looked along the horizon again. 'What else?'

'I've finished *Middlemarch*. That's got a wedding but they make each other miserable. All of them, really. Or there's *Return of the Native*, but it's full of wuthering heath and special effects and we've probably got enough of our own. What about *Waverley*? You'd like that. Walter Scott.'

'What's it about?'

'Masculinity and national identity, mostly. Whether it's better to be Scottish and Romantic or English and reasonable.'

Nothing out to sea, either. The only thing that moves here now is the grass and the water.

'No, what's it about really? What happens?'

She raised an eyebrow.

'Waverley goes to the Highlands and gets involved in several battles and a competition over the girl.'

Catriona rinsed the brush and began to give the pebbles shadows they didn't have.

'Oh, all right then.'

Shortly after one o'clock I saw Ruth working her way down the hillside. The sky was still empty. Catriona had laid out a miniature gallery of watercolours on the padded back of her pack, and was winding braided reeds around a stone. Nina was more than halfway through *Romola*. Ben was pretending to be asleep again and Yianni was composing something that didn't look much like work in his notebook. *Waverley* has the kind of slow-motion plot that needs a captive audience. You'd need an intercontinental flight, probably an intercontinental flight on your own where the movie was in a foreign language, to make the most of *Waverley*. I'd been trying not to think about it, but my stomach rumbled so loudly Catriona looked up from her braiding.

'You'd be wanting your lunch, aye?'

Nina grinned at her. 'Aye, he would that.'

'Shut up,' said Catriona. 'You can't do Scottish.'

'I can.' Nina put her finger on her page and closed the book. She glanced to the North. 'My dad grew up in Edinburgh, I'll have you know.'

Catriona shook her head. 'You never would know. Look, what do you think? I was thinking about a jeweller at home, she makes sort of silver seaweed round shiny stones.'

She held out the chalky pebble in a net of gleaming green reed.

'Ritual or ludic object, significance unknown,' said Nina.

Catriona appraised it.

'I bet we could come up with a ritual.'

'I expect we will, if we stay here much longer.' Nina looked up again. 'Incantation for calling a machine out of the sky. It's going to be easier, isn't it, when we agree it's not coming.'

Yianni threw his notebook on the ground.

'Sorry,' said Nina. 'What about lunch? Jim's hungry.'

'Aren't you?' I asked.

She shrugged. 'It's not very appetising, what we've got.'

Catriona put her pebble down.

'What have we got, exactly? I mean, say the plane doesn't come. How long?'

'It will come.' Yianni picked up the notebook and stuffed it into his briefcase.

'Imagine it didn't. Imagine there's a storm and they have to turn back. Imagine a freak meteorite lands on the runway at Nuuk. How much of a margin have we got? How many days?'

Yianni frowned at his bag, pushing the clip in and out of its socket.

'Catriona, we don't need to have this conversation.' He stood up. 'If you wave to Ruth, I'll get some lunch out.'

It wasn't lunch. It was crackers and water. I looked up and met Catriona's gaze. She looked as if she'd just been hit.

'This bad? This bad already?'

'There's more dry stuff,' I told her. 'I guess there's no point lighting the stove.'

'It's one more thing to put away. They might come any time. We can't keep them waiting.' Yianni scanned the sky again. I feel as if I've got some kind of radar in my head, something sweeping the ether for a hum or a linear movement. Or a plucking hand or signal call. Dear God, let us get out of here.

'There's no one up there,' said Ruth. 'Or in the next valley. The cairn must have been there all along.'

Catriona turned a cracker in her fingers. 'We'd have seen it, Ruth. Wouldn't we?'

By 3 pm we were all reading Victorian fiction. Even Ben attempted some Dickens. By 3.30 the words were blurring on the page and it was too dark to land a small plane on an unlit field. Ruth sighed and put down *Villette*, which Nina had claimed would fit her like a high-heeled shoe. On the cover there was a picture of a girl in a white low-cut gown sitting on an armchair with her head bent over a letter and sausage-like curls hanging down on each side of her face. It lay on the turf.

'Yianni? They're not coming today. It's too dark. We should put our tents back up for the night.'

Catriona made a muffled sound. Yianni looked at the fading sky, and then quickly round to the new cairn. I shivered.

'They'll come tomorrow,' he said. 'They must think we wanted an extra day. Waiting for us to call.'

'No.' Catriona dropped *Middlemarch*. 'No.' Her gaze flicked from one face to the next.

'No, we don't put the tents back up, or no, they won't come tomorrow?' Ruth is calm as if this is how we planned it all along. As if she came here meaning not to go back.

Catriona started to cry. 'No, this isn't happening. No. Just no.'

The sea was beginning to merge into the sky and colour was already leaching out of the hillside above. *No* seemed about right. I couldn't think of anything comforting to say.

Nina moved over to Catriona, her finger still marking her place near the end of *Romola*. She put her arm round her.

Catriona went on crying just the same. Crying doesn't change anything.

'Cat?' said Nina. 'There's nothing we can do about it now. You just – you just have to believe that they're all right. Your family, or whoever. You just have to believe they're getting up and – and going to work and working.' There were tears in her eyes. 'And you have to think of them going home and you're not there. You're not there but they're OK. They'll be OK.' The tears ran but she went on talking. 'And you have to think of them eating and going to bed – without you – and reading and cooking and watching TV and they're all right. And they will be all right. We won't be, but they will.'

Catriona's shoulders shook. Nina hid her face on Catriona's shoulder and went on, her voice rough. 'You have to think, if you were dying, they'd still have a future, wouldn't they? They'd see places you've never been and they'd cook things you've never heard of and they'd meet people you've never known. And you have to think – that they'll do that. You have to.'

Catriona turned and they held each other. The rest of us looked away. The dusk thickened, hiding the cairn above us and the horizon where no plane buzzed. Waves washed the shore.

'We might as well put the tents up here, mightn't we?' asked Ruth.

Yianni shrugged, face averted. 'Saves carrying everything back across the stones.'

He looked at the long boxes we'd been not looking at all day.

'We'll have to put the finds tent back up. In case it rains.'

'Nina?' Catriona rubbed her knuckles into her eyes and looked up, blotchy and damp. 'Do you think – will *they* come here, too? You know.'

She tilted her head towards the hillside and the grave. The plateau was clearer from this side of the river. We'd put the turf back but it wouldn't take root again until spring.

Nina's face was dry but pale. 'We're closer to the chapel.' She paused a moment, listening. 'I hear bad sounds from there. There was someone inside, you know, when it burnt.' She looked over at the ruins. 'I wish there wasn't another night here. I've – I've heard him trying to get out. From the fire.'

'Catriona?' said Ben. 'Be careful. Seriously.'

'OK.' Yianni handed *Jane Eyre* back to Nina. She took it without looking at him. 'We'll get the tents back up. Finds first. I suppose we can just put tarps over the stores. Then our tents. We'll need to be up early again to get ready.'

Catriona and Nina exchanged glances.

'Yianni?'

'What?'

Catriona looked down at *Middlemarch*, on the cover of which a woman in a black dress and a man in a black suit stand by a table in a room with orange curtains.

'How long are we going to keep doing this? I mean, how many days? Before we decide they're not coming.'

'Shut up.' He stood up and stood over her. 'Shut the fuck up.'

I got up and put my hand on him.

'Cool it. OK? She's scared.'

He turned, squared his shoulders, tilted his chin. I can see straight over Yianni's head. The last time I fought anyone was when Bobby Martin said Hannah looked like a squashed red alien when we took her to church that first time. I won then, you might remember. I sat on him and banged his head on the grass until he said what I told him to say and you and Mr Martin came running. I heard you talking about jealousy and new babies to Mrs Martin and then you took me home and locked me in my room for the afternoon and we all had to pray for me to see Jesus in my heart before we could eat dinner. I looked back at Yianni. His nose hung out over his facial hair like a sign swinging outside a shop. Bullseye.

Nina reached out and touched his elbow.

'It's not your fault, you know. No one's blaming you. Yianni?'

'Of course it's my fucking fault, you silly cow. I brought you here. Oh, leave me alone.'

He walked off down the river. The shoreline was dimming and the wind moaned through the grass. You, I thought. That 'you' was singular.

'Is that his problem?' asked Ben. 'He's blaming himself?'

'Makes sense.' Ruth handed *Villette* to Nina. 'Though why men have to threaten women when they're blaming themselves remains a mystery. Come on. I guess we'd better get those tents up.'

Nina took the book. 'It is his fault. He's in charge. And it appears there's no plan B.'

'Tents,' I said. 'Come on.'

I offered Nina my hands and pulled her up. Catriona stayed huddled on her rock. I held out my hands again.

'I almost don't know if I'm more scared of Yianni or the Greenlanders,' she said. 'I keep hoping I'm going to wake up.'

'I'd be more scared of the Greenlanders,' said Nina. She offered her hands as well and between us we pulled Catriona up. They're both light now, lighter than the girls. 'They've got less to lose.'

The longer we stay here, the less any of us have to lose.

Yianni came back when we'd put the tents up in a corral and got the lamp and the stove going. The finds and the stores were well up the hill and even Ruth hadn't objected to that.

'If you're worried about fuel you shouldn't be using the lamp,' he said.

Nina, who was shredding bits of thyme into a pan of water, noodles and garlic, looked up.

'I thought you weren't worrying about saving things?'

I nudged her foot.

'I'm not. You are.' He came and looked into the pan. 'Sorry I was rude, Catriona.'

She was sitting in the darkness behind him, the lantern light flickering on her glasses so I couldn't see her eyes.

'That's OK. I didn't mean to be provocative.'

Nina dropped some more twigs into the water.

'You weren't,' she said.

I kicked her foot again.

'What's for dinner?' asked Ben. We all knew. We'd talked about whether to use the last of the pesto.

'Noodles in garlic consommé,' said Nina. Her voice was

hard. 'Noodles with garlic and fresh seasonal herbs. The cuisine of West Greenland is characterised by a spare, neo-Japanese minimalism. Noodles in a clear broth garnished with locally foraged organic vegetables. Get used to it.'

Catriona pulled her feet closer in to her body.

'Vegetation, more like. Nina, is that grass?'

There was a handful of soft blades in the upturned pan lid at Nina's feet.

'Not the right kind of grass, I'm afraid. D'you suppose the Greenlanders just got drunk in winter?'

'Nina, I'm not eating grass,' said Ben.

'We can't digest it.' Ruth sounded as if she were correcting a student. 'We have no way of processing the nutrients. You'd need an extra stomach. Or two.'

Nina picked up the grass and let it run through her fingers. 'I know. But I bet there's not a lot of nutrition in herbs anyway. I just thought it would add texture. And it tastes a bit sweet. Surely we can process the sugar? I bet you could call it *herbes du Nord* or something and use it to garnish something.'

She put a few blades in her mouth and chewed a moment.

'A mild fish. Freshwater trout. I mean, watercress is traditional but it's too strong a flavour, really. Better with salmon.'

My scalp prickled and I glanced around, but could see nothing outside the small pool of lamplight. The stove and the lamp purred.

'Nina,' I said. 'Stop it. You're creeping me out.'

'One of them's at the grave,' Nina continued. 'And I don't like the feel of the chapel. I've never been this close to it in the dark before. Anyway, I was only going to sprinkle some on each bowl. To make it look better.'

'It's dark,' I said. Dark, and cold. The grass under my boots crunched with frost and my fingers were too stiff to move. When Nina cooks, at least she gets to warm her hands.

She tipped the grass onto the ground.

'OK. Fine. No garnish. No consommé. No fresh seasonal herbs. No *nouvelle cuisine*. Stale noodles with water and garlic. Fewer calories per portion than a piece of toast, and make the most of it because after this it's mostly water, at least for as long as the fuel holds out to melt ice. Enjoy.'

She turned the stove off and began to dump noodles on plates. The pan tipped and she caught it in a gloved hand, but hot water sloshed out.

'Oh fuck. Fuck and fuck. Now I've got fucking burns as well as fucking frostbite.'

Catriona took the pan by its handle.

'Get that glove off and put your hand in the frost.'

Her hand steamed when the glove came off. A blister was already forming.

'First aid kit,' said Yianni.

Nina cradled it in the other hand. 'It'll be all right. I've done worse. Just not in the sodding Arctic.'

She licked the burn.

'Put a dressing on it.' Ruth hadn't moved. 'You'd be stupid to get it infected.'

'OK,' said Nina. 'Jesus.'

Ben passed Yianni the first aid kit. Unzipped, it's frighteningly small. I'd guess Nina's scald is at about the limits of what we're equipped to deal with. The kit you had in the car is bigger. Nina held her hand up while Yianni peeled back her sleeve. He was about to squeeze antiseptic cream onto his finger when Ruth spoke again.

'Wash your hands. Or you're rubbing whatever's on your fingers into blistered skin.'

'I know that. Bloody hell, Ruth.'

He opened a wipe and cleaned his fingers. I watched. He held Nina's bare arm with one hand and spread the cream over her damaged skin with the other.

'You're not looking,' said Ruth. 'Either of you.'

Nina's hair hung down over her face.

'No,' she said. 'Well. As Jim says. It's dark, isn't it?'

I looked away.

'Sit down,' said Yianni.

I looked back to see him kneeling in front of her, her upturned hand resting on his thigh while he unrolled the bandage. They were looking at each other now. You can see in the air, here, when people are breathing fast. Catriona was watching them too. She caught my gaze and turned away to serve dinner.

It was so cold, last night. We went to bed straight after dinner, too cold to sit out in the wind any longer. I wore everything except my coat and boots: shorts, long johns, two pairs of pants, inner and outer socks, two t-shirts, polo neck, two sweaters. Mittens and hat, and the hood of my sleeping bag pulled close round my nose and eyes. The ground was frozen hard under my bones, and after a while my condensed breath on the sleeping bag began to freeze and crackle. My guts twisted with hunger. Hunger hurts, I find. I'm never going to walk past a homeless person again. I lay there. Shaking uses energy, which I haven't got spare, but it generates heat, which I need. I was trying not to think about home, about my bed with Mom's quilt and my stereo with friendly voices and the model airplane and Grandma's rag rug. I want you to know, Dad, if you get this, if I'm not writing into the wind and the dark here, that the time we spent making that plane includes some of the happiest hours of my life. Winter, and the snow swirling outside the garage window, just the fiction of a journey on the path we cleared between the workshop and the kitchen where Mom and the girls were baking the Christmas cookies. Before I'd understood that one day I'd grow up and leave you. Before you started talking about leaving us.

I don't remember Mom making my quilt – it was already there for me when memories began – but I remember her making Holly's and then Hannah's, leaning over the bump to reach the sewing machine and calling me to feel when the baby kicked. The shock of someone's feet inside my Mom. I tried to think about the Greenlanders in their big house up the hill. They knew how to stay warm, with thick stone walls and turf roofs and fireplaces where you could spit-roast an ox, if you had an ox, which they didn't. They spent the winter gathered round the fire, spinning and telling stories and maybe not so much to eat but enough, enough dried fish and dried berries and frozen seal and goose and ptarmigan and grains from the summer harvest to see you through to spring.

Every time I opened my eyes I could still see Nina's torch

glowing. There was a murmur of women's voices, and I closed my eyes again and remembered the women's fellowship nights, when I used to sit at the top of the stairs and smell the different perfumes and the waves of cold air as the door opened and closed. Later, Holly would join me, and she'd wriggle down a few steps on her front and peer through the banisters like someone engaged in urban combat to see what kind of cake was going and if any of the good stuff, the layer cakes and brownies, was going to be left for us. Once I was old enough to stay up and pass the coffee, I used to pass cookies up to Hannah and Holly, and I guess once Holly was old enough she smuggled the less sticky stuff to Hannah. Are you still doing that, Mom? It's the second Thursday today and you're six hours ahead. Maybe right now Hannah's opening the door to Mrs Pearson and taking her coat up to the bed, and Mom's taking the Tupperware box and oh-ing as if she didn't see the pineapple upside-down cake every month for the last twenty years. Though I guess it's more likely Hannah's gone out, off in someone's car with her friends.

I must have slept. I woke to hear crying. Nina's torch had gone off and someone was sobbing, the long, low notes of a person who's been going a long time and has no reason to stop. It could have been Ruth, grieving, or Nina in fear, or Catriona losing hope. It might have been one of the Greenlanders, and it doesn't seem to matter very much anymore. Living or not, being here now seems a good enough reason to cry. The shaking had stopped. My hands throbbed with cold but my feet were still numb. I remembered Uncle Bill telling us about the time Grandpa stayed out too long in the cold ploughing and got his fingers frostbitten, and his hands turned black and swelled up like a corpse's before the nails sloughed off and the skin began to repair itself. And Holly didn't stir a step outside without gloves until May. They must have known how to deal with frostbite, the Greenlanders, with their saunas and fur mitts. One of the outbuildings here looks like a sauna but we've no fuel, of course, and it'll be a cold day here in hell before Yianni lets us burn the turf. Sorry. Not hell, of course. There is still the hope of salvation, a warm heaven, a Valhalla of steaming mead and roaring fires.

I keep trying to pray. The phrases are too worn. I can't find the words anymore. Faith must be stronger than suffering; Christ crucified shows us how the image of God in man can withstand pain. I'm so cold and so hungry and I'm sorry about this but so scared, of cold and hunger and of what cold and hunger might do to us all. It is easy to feel forsaken.

A shot rang out. Always wanted to say that. It's like 'follow that car'. I said that once – well, 'follow that bus', when I missed the Greyhound out of Chicago by about thirty seconds. Apparently they hear it quite a lot, cab drivers. Anyway, it wasn't a shot, of course. If we had guns we could shoot birds, or even a seal. It was fire, the repeating fire of burning wood, and the smell carried too. But with the smell of logs, the smell of the big room on the farm, came the smell of roasting meat, and with the shots ringing out came the soundtrack from the bad moment of a bad movie, the kind of movie the girls should never see. Nina's tent, I thought, as if battery torches could somehow set fire to icy, fire-retardant canvas, but the smell was wood and the voice was male. No wood in Greenland, except for that drifting boat. I sat up but, God forgive me, I was too cold. Too tired. And when I woke again there was silence and darkness and cold.

The plane's not coming. We sat again, through the grey coldness that passes for day. More of the river is frozen and we're running low on fuel. There were crackers, and when we'd finished I went round and picked Nina's crumbs out of the turf and ate them. I meant to finish *Waverley* but I can't concentrate enough for a sentence. It's not terrible. I want you to know that. I'm cold and tired and hungry but as time passes it's not so bad. I'm back with you a lot, back home peeping through the banisters or pausing on the landing to hear the girls' breathing before I go to sleep, sitting at the table watching Mom chopping onions and singing the Battle Hymn of the Republic off-key. When the end comes, it's not going to be too bad, and I want you guys to know that with the last beat of my heart and the last breath in my body I loved you all.

CATRIONA

The plane hasn't come. I hope you were waiting for me. I was thinking about you, cooking the chicken curry that was such a disaster last time, and Rudi going round the market for cheap fruit to make that weird mousse jelly thing his mother sent the recipe for. I bet the candles for the table are exactly where I left them, unless Tibor's been over this summer, and I bet neither of you has chopped an onion since I left. I thought I'd get back and there'd be the smell of coconut in the hall and you'd be in your room, playing poker on the internet and eating cookies your mother posted ten days ago from Canada. Rudi would be working, but when he heard us he'd come out and be irritating, stop you telling me what happened with Tibor. I've no gossip to offer in exchange, I'm afraid. Unless we find a way of using condoms to catch fish, they'll be coming back as intact as I am. But it was a kind thought, and instead I could tell you all about the small-town Midwesterner from Harvard, the grieving New Yorker and Nina, and then when Rudi goes back to his room leaving the washing up you and I could sneak out to LJ's and come home giggly long after he's in bed.

*

Only the plane didn't come. It still might. There's no reason why not. They know where we are, and they must know we're still here. Maybe they just think we want extra time, or they've got the wrong date, or there's been a technical problem and later today we'll hear buzzing in the sky and we'll never tell anyone we thought the end was nigh. I think we all do think that, now. There's a certain relief in it, like the time I should have been doing Highers and got caught truanting in the National Gallery, and all the consequences I'd been dreading for weeks were at last a reality to deal with rather than a worry nagging me through the night. Here we are, at last, the Bomb exploded, the Ice Age under way, the giant asteroid landed, the apocalypse settling into Act II. Plague has swallowed you all up in the night and we're the last people on earth, even though I know perfectly well that epidemics don't work like that. The successful pathogen doesn't kill all its hosts.

The problem is, this dig hasn't been as well planned as you'd hope, and we're already running out of food. Ending with a whimper. This morning we got it all out. There are three packets of crackers, two packets of noodles, a bag of what Nina calls MDF (mixed dried fruit), half a tube of tomato purée, half a box of dried skimmed milk and two packets of angel delight. We'll know we're starving when any combination of the above looks like a good idea. I've got some Kendal mint cake, Nina's got a tin of capers and some chocolate so serious it comes with a handwritten label. Ben has a few M&Ms left in a big packet and the other three either didn't bring any treats, have already eaten them or are still hiding them. It's probably the highest concentration of refined sugar for several hundred miles but it's not going to keep six of us going for very long at all. You could probably eat it in an afternoon while writing up without noticing.

We tried to talk about survival but we didn't get very far. I think postgraduate students may be an evolutionary dead end, though you'd think archaeologists would have more of a clue than mathematical logicians, say, or experts in late Latin poetry. Nina, who works on nineteenth-century travel writing, knows all about the expeditions that didn't make it, but so far her suggestions are limited to wondering how the Franklin survivors cooked their colleagues, since the chopped up bones were found in cooking pots, and suggesting that there ought to be some way of catching fish using tights.

'What, fishnet tights?' I asked. She doesn't look that kind of person, and anyway, surely not here.

'No. Marks and Spencer's best wool. They're very warm.'

'Well, in that case you don't want to use them for fishing, do you? Nutter.'

Then Jim asked what tights were and we all got sidetracked into a discussion about hosiery and transatlantic translation. Jim and Ben said there was no way of distinguishing tights from stockings in American English and we had to wait for Ruth to come back and say 'pantyhose', which Nina said sounded much ruder than tights. It was all preferable to giving any further thought to cannibalism. I do not want to end my days in Nina's pot with thyme and angelica. (It sounds like a folk band, doesn't it, Catriona, Angelica and Thyme? Better a folk band than a recipe.)

Ruth remarked that we'd live a long time on a seal. Ben said he wasn't sure about eating seal, what about trichinosis, and Nina said it wasn't the eating that bothered her, she'd even cook it if someone handed her a nice flat steak.

'We haven't got a gun,' said Jim. (Good.) 'How would you kill it?'

'The Greenlanders managed,' said Nina.

'The Greenlanders had lots of practice,' I pointed out. 'Not to mention a whole summer to get ready.'

We're spending nearly all our time by the tents now. We all looked around for handy weapons.

'I suppose we could make some kind of bow.' Ben sounded as if he was supposing we could make some kind of nuclear-powered robot.

'Out of what, a laptop carrier?' asked Nina.

'A tent pole,' I said. I was rather proud of that.

'I think you'd need to practise,' said Jim. 'I mean, I don't think we can hope to make a bow and arrow out of a tent pole and then just go out and kill a seal with it.'

I caught Nina's eye and started to laugh, and then we all did. It's not funny. I am cold all the time and can't get warm, even at night, even sharing Nina's tent. Cold hurts. The chilblains on my hands are turning dark purple, like bruises, and when I can feel my feet they hurt as if they were forced into boots two sizes too small. My jeans are gaping round the waist, which I still find rather exciting, the thought of all the baklava and chocolate cake and pizza I'll be able to eat later, but I suppose it won't be so exciting in a few weeks. My bones hurt when I sit on the rocks.

Imagine having a bony bum. I hope you're there, reading this. I hope you get the joke.

'Have any of you ever actually killed anything?' asked Nina. 'Apart from an insect?'

'I don't kill insects,' I said. I remembered all those bees in your room when the cherry tree was blossoming.

'What, not at all?' Yianni was trying to sew up a hole in his glove and none of us girls was helping him on principle, although we were all watching and thinking what a mess he was making. Nina had threaded the needle and I told her she was betraying the sisterhood.

'Mosquitoes,' I said.

'I hit a rabbit, once. In the car,' said Ben. He shivered. 'You could feel the bump. On the way back it was still there, with its eye on the road. Horrible.'

'You mean you didn't stop and finish it off?' asked Ruth. Ruth is calm. We need calm. I imagine she could finish off anything that needed it.

'My dad's got this pond, in our garden,' said Jim. 'So you get frogs? And one day I was mowing the lawn, I'd have been about ten –'

I put my hands over my ears, but it doesn't really work. It's like trying not to watch a scary film, sometimes, round here. Except that films end, and you go back out into the street where the need to buy bus tickets and find the keys clears death from your mind.

'Stop it,' I said. 'I don't want to know.'

Yianni jabbed his finger with the needle and sucked it. The glove looks as if it's mutating into a sock.

'So the primal killer instinct isn't exactly raring to go?' said Nina. 'What about limpets?'

We found a lot of limpet shells on the site, but on the other hand we found some butchered dog bones and that doesn't seem like a good idea either, even if we had a dog.

'Maybe they're nice,' she said. 'Maybe it's the next great seafood discovery.'

I could see her thinking about limpet noodle soup or limpet and oatcake porridge. I pulled my sore hands into my sleeves. Limpet sorbet.

'This is survival,' I said. 'Not cookery.'

Nina stopped smiling.

'It's not either at the moment,' she said. 'There's no lunch.'

We spend too much time just sitting, hoping. There's only a few hours of light each day and we come out of our tents clumsy as bears and sit and talk about what we might do. Walk along the coast and find a farmhouse, Ben says, but we know perfectly well we haven't got the energy or the equipment. It might be nicer to die of exposure in a few days than starvation in a few weeks, but dying later seems better than dying sooner. There's always a chance the plane will come. Nina says the history of polar exploration suggests that unless you really know what you're doing, like Shackleton, it's better to stay put and wait for the cavalry, though this, of course, assumes that the cavalry aren't being ravaged by an epidemic, and she did add that it can take a year or two. We talked about building up one of the little rooms in the hall, cutting turf and roofing it in, but Yianni won't hear of 'desecrating' the site, which appears to be sacred to the professional advancement of Yianni Papadatos, not a cause for which the rest of us are particularly eager to martyr ourselves. We talked about trying to light a beacon on the hill top, but once the fuel runs out, which is in a few days, we'll need anything else we can burn to melt ice for drinking. It's not as if we've seen any ships or planes out here anyway. A few plane trails, maybe, at the beginning when the sky was blue and the ground was warm, but nothing low enough to see us. The second day we were waiting, Yianni and Jim laid out stones to make an arrow shape pointing to our tents, but in the morning the stones had been moved, and anyway, I can't imagine that anyone who missed ten tents would see a stone arrow in a field of boulders.

You see, there is someone or something here with us, moving things. It throws stones out of the mist and creeps around at night, moving things. I know this sounds mad. I can imagine you looking up at Rudi and not knowing whether to believe me. (I'm imagining, I realise, that this will somehow be posted and fall into the hall with the takeaway menus and overdue bills. I'm sorry, by the way, about not leaving a cheque for the phone bill. I guess by now you've either lent me twenty quid or been cut off. And I'm sorry about not leaving enough rent. I'm sure my parents will help, if you ask them – get the address from college – but I suppose, again, either you've sorted it out or been evicted. Sorry.) It took me a while to believe in them, these presences or inhabitants or whatever they are. At first I thought it was just Nina. I don't blame you if you don't believe me, I see that like archaeologists, anthropologists can't afford to see ghosts, but Ruth's the only one here who's so thoroughly educated she can discount her own sensory data as culturally produced. I know what I've seen. What I hear.

The snow fell last night. We've had little falls before, an icing-sugar dusting over the rocks, but this is the full Arctic extravaganza experience with special effects, including, at last, the Northern Lights. It's got me painting again. I did some good ones, before it started to be dark so much, and before we had to thaw all the water. They're all in the folder in my rucksack. I thought that was it but I've started again now, sitting in the entrance of the tent until I can't control the brush and then coming back and getting back in my bag until my fingers thaw and then going back out again, as long as it's light. There's something about painting ice with watercolour, the way the same medium is out there as a solid that looks like light, and in the little pot for my brush. Painting water with water, it's like trying to paint glass. I had a go at the Northern Lights but it's a waste of paper and paint. There will be ways of painting electricity in a dark sky with watercolour on A5 – Marguerite Donaldson would be able to do it – but I can't. I know, I'm like the people who went on dancing on the Titanic or whatever it was, but one might as well. We all are, with our jokes about food. Dancing on the way down. Laughter and art are forms of defiance, a way of staying human. And it's not as if I'd otherwise be out there strangling polar bears with my bare hands and dragging them home for Nina to casserole. It passes the time and if we do get out of here, I'll have some great paintings. Like being Keats or Sylvia

Plath but then surviving to rake in the glory, Nina says, but don't rely on my paintings for the rent. (You couldn't rely on Plath or Keats for the rent either, Nina adds. Or in fact any really good poet she can think of, except possibly the later Wordsworth, who is less good than the earlier and less reliable Wordsworth. Now you know about Nina.)

Nina's here. We've been sharing a tent since the plane didn't come. I can't imagine what it's like for the others, spending so many hours entombed in the dark on their own, but Nina's got torch batteries and books as well as useful observations about the relationship between poetic genius and financial reliability. I'm not going to die bored, as long as Nina's around, and that must count for something.

Ben and Yianni had a row. We've avoided most of that so far, managed to behave as if we're in the Graduate Common Room and the biscuits are running low, but hunger, I find, eats your mind after a while. Nina's stories are less funny than they were. We are probably not all going to get out of here. Someone will go first. In my more broad-minded moments, I wouldn't mind being eaten as long as I was already dead of something other than someone else's wish to eat me, but I don't want to watch us all getting scared of each other. Ben wants to move into the farmhouse, where at least we'd have shelter and the basis of some insulation, not to mention a more-or-less functional fireplace, and Yianni says he won't be remembered as the guy who trashed his own site. We move up there, he says, over his dead body. Ben said Yianni should realise that there will come a time when that price is worth paying. It took me a minute, looking into Nina's pale face. A threat to kill. Ruth's right, you know, it's daft to fear the dead when the living are so patently terrifying. It's not, anymore, the ghosts I'm worried about. (You never were, says Ruth. You were just projecting your worst fears onto something less scary than life. Piss off with the cod psychology, says Nina, it's not my worst fears that built the cairn. I distract them both by wondering if cod stocks have recovered to the point where there might be some out there, if we could think of a way of catching them.)

Jim said we should vote on it, whether to move up to the house, and Ruth said questions of survival were no place for democracy and Nina said Ruth had just encapsulated the limitations of American politics and Yianni said Nina for fuck's sake shut the fuck up and I sat there shaking and thinking I'd be sick if I'd had anything to throw up. I felt like that before I phoned up for my Finals result, worse as the day neared, but I was wrong. There was the possibility, actually the probability, of a good outcome. Uncertainty is our last luxury and it's running out.

Ben's right, though. Moving onto the site is probably our best hope. We didn't vote, which is at least in the short term a good thing because enforcing the democratic result would be categorically disastrous, but if we did – when we do – my money's on the Greenlanders. We don't have their skills, much less their stores of meat and fish and dried berries, but at least they left us their house. We don't know how to put the roof back on, but surely knowing how to unbuild is a good start? (Deconstruction as a guide to survival, muses

Nina. It's horribly probable as the end of human evolution, wouldn't you say?) Enough. I must stop this.

There's one more thing I want to say to you. It's why I'm writing to you now and not my parents. I love you. I love the way you look up through your hair when you laugh, and I love the way you laugh at your awful cooking. When you went home last summer I got no work done and wandered round the Botanic Gardens thinking of you and composing e-mails I could never send. Sometimes I went home and typed them out, careful to leave the 'to' field empty so I couldn't send one by mistake. When you came back and hugged me I wanted to kiss you, the first want of my life, and all this year I've been wanting. I think of you now all the time, the rise of your eyebrows, the curve of your shoulders in that vest top. The way the sun shone on you at my party. You've got Tibor, I know. You love him, sort of, inasmuch as you can love someone without their consent. I won't bother you with this. You won't even see this, probably. But it would have been too sad, to leave it unsaid.

YIANNI

I've screwed up. I'm sorry. This is my fault. That's all I want to say, really. I'm sorry to you. I'm sorry to my family. Most of all, I'm sorry to the others. And their families. To David, for taking Nina away from him. Nothing I can say is going to help, is it? This is like one of those dreams where you've done something so terrible it can only be the manifestation of the evil person you always suspected yourself to be. You wake up feeling dirty, feeling as if the crime is the proof of your secret awfulness because even if you didn't actually do it (this time), your mind thought of it, planned it, executed it. Executed her. It's only good luck you were asleep at the time.

Last words. I might as well. It's not as if I'm going to know when you've read this. In the night I thought I killed her. Came on her down by the river, breaking ice for water, dragged her over the snow by those chicken-bone arms, surprisingly heavy, and her boots digging into the turf as she twisted around. Hit her face, again. And then again. And then what I cannot say. Her white legs beating like wings on the snow. Not even writing it down would make me realise *I didn't do it*. And then she was crying, the way she never stopped crying, so I took a rock and banged it into her head until she did stop.

After that, it was the clearing up. Blood on the rock, of course, and on the snow. Blood and worse. I threw the rock into the river, well out, into the middle, and it made a crash and then a gulping sound as it hit the water. There's nowhere to hide a body here. The river is frozen, the beach too exposed, the snow not deep enough. If I could have got her up to the grave it would have been a good place, at least a poetic place, but we filled it in when we were first planning to leave. Later, maybe, I could move her up there. In the end I bent her double under a boulder and rolled two more over the body. The boulders left bare patches and hair stuck out, if you looked closely, but who's going to look closely?

I did not do this. I did not do it. Absolve me.

I think we are all going to die here. We are, in fact, dying, and that is my fault. I have killed, will have killed, five people. There's nothing I can do about it. I'm almost certainly going to have to watch at least some of them die. Cope with the bodies. I never imagined it would come to this. It wasn't in my plans. Research fellowship. Monograph. Major funding bid. Criminal negligence. Death by starvation and exposure. Nina can't lift a five litre water bottle now. There are black patches on Catriona's feet and fingers. Jim lies in his bag talking to his parents, who are several thousand miles away and probably dead. I caught Ben scraping lichen off the rocks to chew and you can see the bones moving in his face when he talks. I am the leader. Maybe we should have tried to walk along the coast, the way the shepherds went, but we don't really know where they went. You know where to look for us here, don't you, if you do decide to look? You know where we are. That's why I thought we should stay.

Technically, I think this has been a success. The notes are on my laptop, which seems to be still functioning. My password is Marielen1973, and the folder is 'Greenland site notes'. There's another copy on the red memory stick in the zipped pocket of the bag and another in the black one in the inner pocket of my jacket. I was planning to try *Acta Archaeologica* with the findings from the grave; James Richardson's planning a special issue on medieval battlefield burials for next fall and I thought this would be an interesting angle. I've uploaded the photos. I think the second three under 'burial Sept week 1' show the injuries most clearly.

I've followed the BAA procedures on storage as far as possible. Most of the finds will be OK through the winter, barring disturbance. I imagine that if bears come, they'll eat us before they bother with the burials, but I can't see how to secure anything properly with cardboard boxes and tents. I've done my best.

Some of the others want to move onto the site. I don't know who's going to go first here, but I want you to know that if they have done that, and especially if they've damaged it in any way or lit fires, it was only when I was physically incapable of preventing it. Nina says it's an archaeologists' version of the dilemma of the old lady in the burning art gallery, save the archaeologists or save the archaeology, but she's wrong. (Nina adds a typical aside about why it has to be a lady and why old ladies are thought incapable of saving themselves, not to mention saving the *Mona Lisa*.) The point about the burning art gallery is that you can choose between art and life, but if I made that choice I did it inadvertently when I didn't check the phone before we needed it. (Of all I've done wrong, I think I feel worst about that. I didn't even open the box until we'd lost the internet; for all I know the handset is missing some vital part that got left in Christine's cupboard back at the department.) I didn't check. Didn't blow the budget flying in more food than I thought we'd need. Didn't bring a VHF as back-up. The old lady was history weeks ago. We are not going to survive this, but I don't want my professional legacy to be the destruction of the site I was given £100,000 of ESRC money to explore. I'm sorry. I wish it had not come to this. Since it has, I hope you find everything in order. I hope the research grant, at least, has not been wasted.

Give my love to everyone. You've been my English family, and I'm so grateful for your guidance and friendship these ten years. I'm still remembering how you and Helen bailed me out that summer, and those days in your garden while I sorted myself out. If I've achieved anything, it's been because of your support, and I hope you believe that I never meant to let you down like this in the end.

BEN

Nina thinks we should all leave letters. I'm not sure there's much to say. I'm not writing to say goodbye. Not yet. If it comes to it, I'll die walking, on my way home. For now it's better to stay here. We've moved up to the hall, all except Yianni. Nina used to see ghosts up here. I suppose if you're going to see ghosts, it's a good place for it. She says they've gone quiet since the light went. In the sagas, even ghosts are more active in summer. Ghosts or not, I'd have been up here the day after the plane should have come. I'd have mended the roof and rebuilt the fireplace, worked out how to get a seal and maybe some fish before the bay froze. I regret now that I didn't insist. Me and Ruth made a stash of turfs before the snow fell, before the big row. Yianni asked if we knew how long it takes peat to recover. Yeah, I said, longer than it takes humans to die of hypothermia, and when it comes to a choice I prefer people to turf. And to archaeological sites. And to the pristine status of the West Greenlandic coastline. So now he's down there in his tent, brooding over the artefacts, and we're up here. At least in body. Jim talks a lot, though not to us. He seems to be in Deer Creek living out an

American fantasy with his Mom and his Pop and his sisters and the dog. Sometimes his Nan seems to be there too. He's happy, anyway. Getting weaker but happy. As anthropologists suggest, there are times when it's an evolutionary advantage to be short. We've stopped asking Jim to do much, and to be fair he's stopped asking for much. It makes sense to spend what we've got on the people who can get more. He gave Nina the last of his limpets without being asked. She'd been down on the beach prizing them off the rocks all the time it was light. All two hours of it. In some ways, it's probably no bad thing Yianni's staying down there. It's only two hundred feet away but two hundred feet feels quite far now. The more I realise how he planned, or didn't plan, this expedition, the less time I want to spend with him. He brought the kind of first aid kit you'd expect in a primary school. No glucose tablets, we found, when Catriona went all pale and shaky the last time she was down on the beach. The beach is the only place to look for food, now, and we found some. A dead gull. One good thing about the cold is that dead fauna doesn't rot. I don't know what it died of but I know what we'll die of if we're picky. Nina says we can boil it with seaweed. She says people have eaten worse, and we peer down to see if Yianni's torch batteries are holding out. I would eat worse, if I had to. I wouldn't kill but I'd eat. We're not as good about taking his portion down as we were. More often than not, taking it down there would use more of our energy than he'd get from eating it. Nina visits most days.

*

We keep planning meals. Do you remember that Death Row recipe book Liz sent? And we wondered how anyone on Death Row could think about food? Well, you think about food because it's the only thing left, and now we're up on the site and everyone's talking again we talk about it too. It's better up here. We're colder and hungrier, of course. I saw Ruth stop and rest while she was scraping snow into the bucket for water today. But we get a turf fire going in the old fireplace, we have a hot drink that we call soup once a day and we keep bottles of water in our sleeping bags so it doesn't freeze, and we lie around in the tents. Now there's no risk of rain, I've stuffed the space between the inner and outer layers with grass for insulation. We found some big round stones at floor level. We couldn't work out what they were doing inside the house, but Ruth's started to put them inside the fireplace and when the fire dies down we wrap them in our towels and take them into our sleeping bags. You should try it, next winter. They get hotter and stay warmer than rubber hot water bottles, and no leaks. I've got that beach towel Liz brought back from Cyprus, the one with the dolphin on that you always hated, Mum. It's worn through. I reckoned it was big and you wouldn't miss it, and it's funny now, remembering the towels at home. Blue for Dad, green for Mum. Liz's purple and my brown. Hand towels, body towels, bath towels and facecloths. I'm missing that almost as much as food. A hot shower, a rough towel. A clean shave and an ironed shirt. Louise would laugh to see me now, with a beard. Maybe.

The others just talk about food. It was always Nina's thing, but now everyone joins in, even Jim. Catriona wants Lebanese from a place in Edinburgh she goes to with her housemate. Jim said roast lamb, last time he was talking to us, and Ruth said what I was thinking. He could have had it if we'd known what was coming. Or not coming. We could still have had it if we'd used the old meat-hooks in the chimney or the old cellar in the next room. Ruth wants cheese from a particular shop in Paris and bread people used to queue round the block for, and then Nina said even most French bakeries buy in the dough now. I know, said Ruth, but the one on the rue St Catherine doesn't and that's why people like it, OK? Here we are getting by on melted snow and hot stones and that pair are fighting over bakeries in Paris.

I suppose your trip to Paris got cancelled, with the epidemic. I try not to suppose much else, when I can help it. I think about the pair of you on the allotment, Dad weeding the vegetables and Mum in that deckchair with the Sunday paper as if it's the Mediterranean and not the M1 behind the fence. As October goes by I think about Bonfire Night, the way we used fuel just for the hell of it, just for the fun of watching things burn, the way we handed out sweets as if energy was free. I think about Louise up on the moors with the sky grey behind her, crossing the becks in her old boots, and I know she'll be OK. And if I'm the lucky one out here, with the limpets and the dead gull, I don't want to know just now.

I'm not saying goodbye. I'm still here. And you are too, aren't you, whoever you are, reading this?

NINA

When I came home from the farmers' market today there was a thin blue letter among the brown envelopes and takeaway leaflets on the mat. Airmail paper seems as if it could only come from the past, an era before e-mail when people tried to save a few pence on a stamp, and anyway it's not the time of year for handwritten letters. I picked it up, feeling like someone finding wartime ordnance on the beach, saw the boxy writing shaped by a different alphabet and felt my heart lurch. I put it back, as carefully as if the unusual shape under the seaweed had indeed turned out to be a landmine, and took the bags into the kitchen. I'd been struggling with the shopping, having remembered to take a canvas bag (free with my renewed subscription to *Gastronomica*) and then bought so much I'd had to demean myself with plastic as well, one in surgical blue, now beaded with the blood from a shoulder of marsh lamb which banged its cold flesh against my leg all the way home, and one from a French bookshop left in the Foragers' box of reusables by someone with cultural capital to spare. The handles had made welts in my fingers and my hair was stuck to the sweat on my forehead, so I had some excuse for leaving the letter while

I unpacked the shopping and washed my hands and face. I put the kettle on and found that blue mug you gave me. I emptied, rinsed and refilled the tea-strainer and stood there. The curtains were still drawn in the opposite flat. The kettle was still gathering steam. I went into the hall. The envelope lay there, unexploded. I picked it up again and deciphered the return address. Crete, not Athens, even though it's only May. Your dad must still be off work. The kettle boiled and I took the letter back to the kitchen, poured the water, thought about a biscuit and then washed a handful of early cherries instead. She'd used several sheets of paper, more than anyone would need for the invitation we've been expecting. I decided to marinate the lamb after all and went out to the balcony for rosemary, leaving the letter by the kettle.

It's been a day for procrastinating, one of the days allotted to waiting for the balance of my mind to swing from reluctance to start writing the conclusion to discomfort at carrying it around unwritten. I find myself increasingly reaching for metaphors of excretion when thinking about my thesis, but I have to admit that my alternatives were nearly exhausted. One reaches a point where only unforeseen disaster could legitimate further postponement, and I suppose from that point of view your mother's blue envelope brought a kind of reprieve. I did all right this morning, at procrastination. I stayed in bed until the *Today* programme was replaced by people who ought to know better discussing books and plays about which they know little, ate some stewed apples and yoghurt, dressed and left the flat, wandered round the market spending David's money and had a long chat with the butcher about whether they'll be allowed to sell game again in the autumn. He thinks not, thinks that though all the evidence is that stocks are recovering there's no way of systematically testing wild birds or even deer. I do worry that all this seems to be sending the food scene back about twenty years. People would rather have things sterilised in tins or frozen or hydroponically grown in glass boxes than risk ingesting a bacterium that tastes of sun or soil or, in the case of the lamb, salty marsh grass. I'd say it's a pretty clear sign that a person needs to get back to her thesis when the butcher gets bored of chatting about meat before she's ready to move on, wouldn't you? I went to the Foragers' stall for some of the purple seaweedy stuff they had last spring. I cooked it with salmon then, which was fine, but I had a feeling it might go rather well with the lamb and making the Foragers feel loved and wanted seems almost as important as relieving myself of the conclusion. Ollie's not around any more, I think. I haven't seen him since I got back and I haven't asked, but Jason was there and so was the purple seaweed and we talked – well, I talked – about the bit in *King Lear* with the one who gathers samphire and he said there's no need to hang off cliffs because you can pick it off rocks on the beach like mussels and I remembered the mussel shells under the wall and felt cold, though it's one of those days when even in London the wind smells of leaves, and I put the winter coats away during a preliminary bout of procrastination last week. I bought some rhubarb from the macrobiotic people (harvested at full moon under the sign of Jupiter or something, but better to subsidise nonsense than pesticide and air freighting) and by then it was all so heavy I really had to come home. And find the letter.

When the lamb was anointed with olive oil and honey and rosemary and garlic I washed up. I even dried and put everything away. I turned the letter back over. She'd have written it at the kitchen table, with the icon behind her and the olive tree tapping at the window and the sea Aegean blue in the background. It would have taken a while, in English and on that blue ricepaper, and then she'd have folded all the sheets and licked the envelope and maybe walked down to the village with it straight away, before she changed her mind. I pushed it away and began to make a rhubarb and polenta cake with a honey syrup, leaving an exclusion zone around the envelope as if it were leaking toxins. When the cake was in the oven I washed up again and sat down. It was past lunchtime, almost time for *The Archers*, and usually I'd have made a sandwich and caught up with Ambridge, but I didn't feel much like eating. The smell of vanilla began to drift from the oven. The letter lay on the table. I reached out and opened it.

At first I felt sick. My hands shook as I turned the tap on, refilled the blue mug with cold water. I leant against the counter. There was a child in the yard below, walking up and down bouncing a red ball as if her bouncing were keeping us all alive. She's usually here at weekends, staying with her dad on the other staircase. There used to be a little brother as well. She nearly missed a beat, lunged sideways and kept going. The slap of ball on tarmac bounced off the walls, amplified. We promised we'd never go back, you know. All of us, when we said our final goodbyes at Nuuk. Cat and I sat together on the plane back to Copenhagen but Ben was barely talking to me by then and the last I saw of him was the back of his head as he elbowed his way out into the arrivals hall. I guess the risk of meeting either of the Americans again must be small, and in some ways the knowledge that Jim is out there in the world is comforting, but British academia being what it is the likelihood of bumping into Ben again, back home and at someone's launch party or by the lockers in the BL, is horribly high. It's like knowing you committed a crime and living in fear of meeting a witness but I didn't, did I? I brought you food until the storm came and I couldn't walk any more. If there's a criminal, it's Ben, the only one who was still mobile when we heard the boat. He went straight past you on his way down to the beach.

I finished the water and wandered into the other room. Dust drifted in the sunlight. My laptop gazed from the desk like some kind of surveillance camera run by my superego so I went out onto the balcony with my id, where the noise of traffic and the light glinting off the leaves of the plane tree across the road offered refuge from both the laptop and the letter. David's wanted to see the dig all along and I know he thinks I should stand there again with him, incorporate what happened into my life. Our life. We haven't talked about the Greenlanders. They went away in the end, when the snow came and the river froze, and I'm not quite sure, from here, from my balcony with its view of the bus stop and Sam's newsagent, through the prism of the Selective Serotonin Reuptake Inhibitor I swallow with the first cup of Earl Grey every morning, if they were there or not. I'm not quite sure I want to find out.

I took a deep breath of London air, the sort that makes you sneeze in black, and leant over the railing. Wherefore art thou, Romeo? (Valuing paintings in Chalfont St. Giles, that's where.) There was a queue at the bus stop, an old lady with a shopping trolley and a mum with a pushchair and a couple of disaffected youth with wires coming out of their ears, just the same as before. And mostly it is like that, you know, in London. I'd always wondered how Virginia Woolf could be so flippant about the 1918 Spanish flu in her journal, slipping it in as a joke between Lytton Strachey's sore finger and Lady Murray's invitation to lunch, when the death rate in parts of London was higher than it had been in the trenches and people who had been well at breakfast were dead by bedtime and deadly as plutonium to everyone who saw them in between, but I think I understand it now. When you're not dead, life goes on and there are buses to catch and lamb to cook. Doctoral theses to write. And letters to read, and answer. I thought I would phone Cat, who has given up her doctorate and is painting in Skye, and see what she thought of your mother's request.

Cat said, predictably, that it was an insane idea, but I never claimed to be sane, did I? So here I am. The river is still wavering over the stones and the dark screes where the watcher waited are unchanged. The sky is grey today and so low over the black sea that everything's horizontal and it feels as if there's not much space. The flowers are different from the ones that were out when we arrived last year, fragile white bells that remind me of porcelain cups so thin you can see your fingers through them. There are so many I can't help treading on them. Things came true for you here, your mother says. This is the place that makes sense of your life, by which I think she means your work, or maybe your death. Despite the little pills, I still see the valley in my dreams, hear the wind whistling over your tent and fight my way through snow to reach you because in my dreams it is not too late. To dream of cold and wake to a down duvet and the cotton-scented warmth of your beloved is a pleasure even greater than dreaming of crime and waking to innocence. In some ways Ruth was right, you know, about what haunts the mind. Even our fucked-up and terrifying reality is only very rarely as bad as our dreams. Though you, of course, like her, are one of the exceptions, and I wouldn't want you to think I don't dream about that too, about those final days when you lay alone in the dark as if no one cared when you took your last breath. Hypothermia, Dr Jackson assures me, is among the nicer ways to leave, especially when the brain is already out to sea with hunger. I hope someone has told your mother.

The sea is empty as usual. The grass has grown back over the robbed graves, though the traces of our fire still stain the stones of the house. I thought of cleaning them off for you, engaging in some kind of reconstructive archaeology, but I'd only leave some other trace, wouldn't I? Even hiding traces leaves traces. And we were there. We are history too. There's a boat bobbing by the rocks again, but this time it has an outboard motor and the bright figures down on the beach are David and Nils the pilot, who will take us back to the farmhouse for a supper which will be, happily, neither authentic nor, in essentials, based on local ingredients. I am not stupid. I have no intention of spending a night here. But I sense no presences now, not even yours, and it's time to go.

I open the box, empty your ashes onto the wind and watch them drift and settle like dark snow on the pale flowers of West Greenland.

ACKNOWLEDGMENTS

This book has been shaped by the readers of which writers dream. Thank you to Anna Webber at United Agents, who realised its potential in so many ways, to Amber Dowell, and especially to my editor, Sara Holloway, who always turned out to be right. Scarlett Thomas encouraged me from before the beginning. Caroline Dawnay, Sarah Ballard and Hannah Griffiths offered invaluable support and advice at an early stage. I am grateful for Katharine MacDonald's expert opinions as well as the gift of her friendship.

I thank my colleagues in the School of English at the University of Kent, especially Scarlett Thomas, Jennie Batchelor and Rod Edmond, for moral support. The final drafts of this book were made while on study leave from the University in Autumn 2008 and I am particularly grateful to Malcolm Andrews for protecting this time at some cost to himself. Early research took place under an Arts and Humanities Research Board doctoral grant and a Blaschko scholarship from Linacre College, Oxford.

When mothers write, there are people to thank. Thank you to Hannah Ludlow for being the perfect nanny, to

Georgina Maude for repeatedly bailing us out at short notice, to Sharon Dixon, Sinead Mooney, Adelina Comas-Herrera, Hilde Hagerup, Diane Houston and Jennie Batchelor for fellowship, and to Anthony Maude, as always, for everything.